# Everything in Its Place

# Everything in Its Place

## First Loves and Last Tales

OLIVER SACKS

PICADOR

First published 2019 by Alfred A. Knopf,
a division of Penguin Random House LLC, New York,
and in Canada by Alfred A. Knopf Canada, a division of
Penguin Random House Canada Limited, Toronto

First published in the UK 2019 by Picador
an imprint of Pan Macmillan
20 New Wharf Road, London N1 9RR
Associated companies throughout the world
www.panmacmillan.com

ISBN 978-1-5098-2179-2

www.oliversacks.com

Owing to limitations of space, permissions and
previously published credits appear on pages 265–267.

1 3 5 7 9 8 6 4 2

A CIP catalogue record for this book is available from the British Library.

Printed and bound by CPI Group (UK) Ltd, Croydon, CR0 4YY

Visit **www.picador.com** to read more about all our books
and to buy them. You will also find features, author interviews and
news of any author events, and you can sign up for e-newsletters
so that you're always first to hear about our new releases.

# Contents

# Contents

# First Loves

# Water Babies

We were all water babies, my three brothers and I. Our father, who was a swimming champ (he won the fifteen-mile race off the Isle of Wight three years in succession) and loved swimming more than anything else, introduced each of us to the water when we were scarcely a week old. Swimming is instinctive at this age, so, for better or worse, we never "learned" to swim.

I was reminded of this when I visited the Caroline Islands, in Micronesia, where I saw even toddlers diving fearlessly into the lagoons and swimming, typically, with a sort of dog paddle. Everyone there swims, nobody is "unable" to swim, and the islanders' swimming skills are superb. Magellan and other navigators reaching Micronesia in the sixteenth century were astounded at such skills and, seeing the islanders swim and dive, bounding from wave to wave, could not help comparing them to dolphins. The children, in particular, were so at home in the water that they appeared, in the words of one explorer, "more like fish than human beings." (It was from the Pacific Islanders that, early in the twentieth century, we Westerners learned the crawl, the beautiful, powerful ocean stroke that they had perfected—so much better, so much more fitted to the human form than the froglike breaststroke chiefly used until that time.)

For myself, I have no memory of being taught to swim; I learned my strokes, I think, by swimming with my father—though the slow, measured, mile-eating stroke he had (he was a powerful man who weighed nearly eighteen stone) was not entirely suited to a little boy. But I could see how my old man, huge and cumbersome on land, became transformed—graceful, like a porpoise—in the water; and I, self-conscious, nervous, and also rather clumsy, found the same delicious transformation in myself, found a new being, a new mode of being, in the water. I have a vivid memory of a summer holiday at the seaside in England the month after my fifth birthday, when I ran into my parents' room and tugged at the great whalelike bulk of my father. "Come on, Dad!" I said. "Let's come for a swim." He turned over slowly and opened one eye. "What do you mean, waking an old man of forty-three like this at six in the morning?" Now that my father is dead, and I am almost twice the age he was then, this memory of so long ago tugs at me, makes me equally want to laugh and cry.

Adolescence was a bad time. I developed a strange skin disease: "erythema annulare centrifugum," said one expert; "erythema gyratum perstans," said another—fine, rolling, orotund words, but neither of the experts could do anything, and I was covered in weeping sores. Looking, or at least feeling, like a leper, I dared not strip at a beach or pool, and could only occasionally, if I was lucky, find a remote lake or tarn.

At Oxford, my skin suddenly cleared, and the sense of relief was so intense that I wanted to swim nude, to feel the water streaming over every part of me without hindrance. Sometimes

I would go swimming at Parson's Pleasure, a bend of the Cherwell, a preserve since the 1680s or earlier for nude bathing, and peopled, one felt, by the ghosts of Swinburne and Clough. On summer afternoons, I would take a punt on the Cherwell, find a secluded place to moor it, and then swim lazily for the rest of the day. Sometimes at night I would go for long runs on the towpath by the Isis, past Iffley Lock, far beyond the confines of the city. And then I would dive in and swim in the river, till it and I seemed to flow together, become one.

Swimming became a dominant passion at Oxford, and after this there was no going back. When I came to New York, in the mid-1960s, I started to swim at Orchard Beach in the Bronx, and would sometimes make the circuit of City Island—a swim that took me several hours. This, indeed, is how I found the house I lived in for twenty years: I had stopped about halfway around to look at a charming gazebo by the water's edge, got out and strolled up the street, saw a little red house for sale, was shown round it (still dripping) by the puzzled owners, walked along to the real estate agent and convinced her of my interest (she was not used to customers in swim trunks), reentered the water on the other side of the island, and swam back to Orchard Beach, having acquired a house in midswim.

I tended to swim outside—I was hardier then—from April through November, but would swim at the local Y in the winter. In 1976–77, I was named Top Distance Swimmer at the Mount Vernon Y, in Westchester: I swam five hundred lengths—six miles—in the contest and would have continued, but the judges said, "Enough! Please go home."

One might think that five hundred lengths would be monotonous, boring, but I have never found swimming monotonous or boring. Swimming gives me a sort of joy, a sense of well-being so extreme that it becomes at times a sort of ecstasy. There is a total engagement in the act of swimming, in each stroke, and at the same time the mind can float free, become spellbound, in a state like a trance. I have never known anything so powerfully, so healthily euphoriant—and I am addicted to it, fretful when I cannot swim.

Duns Scotus, in the thirteenth century, spoke of *"condelectari sibi,"* the will finding delight in its own exercise; and Mihaly Csikszentmihalyi, in our own time, speaks about "flow." There is an essential rightness about swimming, as about all such flowing and, so to speak, *musical* activities. And then there is the wonder of buoyancy, of being suspended in this thick, transparent medium that supports and embraces us. One can move in water, play with it, in a way that has no analogue in the air. One can explore its dynamics, its flow, this way and that; one can move one's hands like propellers or direct them like little rudders; one can become a little hydroplane or submarine, investigating the physics of flow with one's own body.

And, beyond this, there is all the symbolism of swimming—its imaginative resonances, its mythic potentials.

My father called swimming "the elixir of life," and certainly it seemed to be so for him: he swam daily, slowing down only slightly with time, until the grand age of ninety-four. I hope I can follow him, and swim till I die.

# Remembering South Kensington

I have loved museums as far back as I can remember. They have played a central role in my life in stimulating the imagination and showing me the order of the world in vivid, concrete form, but in a tidy form, in miniature. I love botanical gardens and zoos for the same reason: they show one nature, but nature classified, the taxonomy of life. Books are not real in this sense; they are only words. Museums provide arrangements of the real, exemplars of nature.

The four grand South Kensington museums—all within the same plot of land and all built in the same High Victorian baroque style—were conceived as a single, many-aspected unity, a way of making natural history and science and the study of human cultures public and accessible to everybody.

The South Ken museums (along with the Royal Institution and its popular Christmas Lectures) were a unique Victorian educational institution, and they still represent for me, as they did in childhood, the essence of museumhood.

There was the Natural History Museum, the Geology Museum, the Science Museum, and the Victoria and Albert

Museum, devoted to cultural history. I was a science type and never went to the V&A, but the other three I regarded as a single museum and I went to them constantly, on free afternoons, on weekends, on holidays, whenever I could. I resented being shut out of them when they were closed, and one night I contrived to stay in the Natural History Museum, hiding myself at closing time in the Fossil Invertebrate Gallery (not as well guarded as the Dinosaur Gallery or the Whales) and spending an enchanted night alone in the museum, wandering from gallery to gallery with a flashlight. Familiar animals became fearful, uncanny, as I prowled that night, their faces suddenly looming out of the darkness or hovering ghostlike at the periphery of the flashlight. The museum, lightless, was a place of delirium, and I was not wholly sorry when morning came.

I had many friends in the Natural History Museum—*Cacops* and *Eryops,* giant fossil amphibians whose skulls featured a hole for a third, pineal eye; the cubomedusan jellyfish *Charybdea,* the lowliest animal with nerve ganglia and eyes; the beautiful blown-glass models of *Radiolaria* and *Heliozoa*—but my deepest love, my special passion, was for the cephalopods, of which there was a magnificent collection.

I would spend hours looking at the squids: *Sthenoteuthis caroli,* stranded on the coast of Yorkshire in 1925, or the exotic, soot-black *Vampyroteuthis* (only a wax model here, alas), a rare abyssal form with an umbrella-like web between the tentacles, spangled with brilliant, luminous stars in its folds. And, of course: *Architeuthis,* the emperor of giant squids, locked in mortal embrace with a whale.

But it was not just the giants, the exotica, that held my attention. I loved, especially in the insect and mollusk galleries, to open the study drawers beneath the cases to see all the varieties, the markings, of a single species or shell, and how each variety had its own, favored geographical location. I could not, like Darwin, go to the Galápagos and compare finches on every island, but I could do the next best thing in the museum. I could be a vicarious naturalist, an imaginary traveler with a ticket to the whole world, without leaving South Kensington.

And sometimes, after the museum staff got to know me, I would be let through a massive locked door into the private realm of the new Spirit Building, where the real work of the museum was done: receiving and sorting specimens from all over the world, examining them, dissecting them, identifying new species—and sometimes preparing them for special exhibits. (One such was the coelacanth, the newly discovered "living fossil" fish *Latimeria,* a creature supposed extinct since the Cretaceous.) I spent days on end in the Spirit Building before going up to Oxford; my friend Eric Korn spent an entire year there. We were all in love with taxonomy in those days—we were Victorian naturalists at heart.

I loved the old-fashioned glass-and-mahogany look of the museum and was furious when, in my university days in the 1950s, it got all modern and gaudy and started installing trendy exhibits. (It eventually went interactive.) Another friend, Jonathan Miller, shared my disgust as well as my nostalgia. "I have a great hankering for that sepia-tinted era," he once wrote to me. "I long endlessly for the whole place suddenly to be plunged into the gritty monochrome of 1876."

Outside the Natural History Museum was a pleasant garden,

presided over by trunks of *Sigillaria,* a long-extinct fossil tree, and a miscellany of *Calamites.* I was drawn to this, to fossil botany, with an almost painful intensity; if Jonathan was nostalgic for the gritty monochrome of 1876, I wanted the green monochrome, the fern and cycad forests of the Jurassic. I even dreamed at night, as an adolescent, of giant woody club mosses and tree horsetails, primeval giant gymnosperm forests enveloping the globe—and would wake furious to think that they had long since disappeared, the world taken over by brightly colored, up-to-date modern flowering plants.

From the Jurassic fossil garden of the Natural History Museum it was scarcely a hundred yards to the Geology Museum, a museum virtually deserted at all times, as far as I could see. (Sadly, this museum no longer exists; its collection has been incorporated into the Natural History Museum.) It was full of special treasures, secret pleasures, for the knowing, patient eye. There was a giant crystal of antimony sulfide, stibnite, from Japan. It stood six feet high, a crystalline phallus, a totem, and it fascinated me in a peculiar, almost reverential way. There was phonolite, a sonorous mineral from Devils Tower in Wyoming; the keepers of the museum, once they got to know me, would let me strike it with the palm of my hand, and it would emit a dull but gonglike and reverberant boom, as if one had hit the sounding board of a piano.

I loved the sense here of a nonliving world—the beauty of crystals, the sense that they were built of identical atomic lattices, perfect. But if they were perfect, mathematics incarnate, they also stirred me with their sensuous beauty. I spent hours studying pale yellow crystals of sulfur and mauve crystals of fluorite—clustered, gemlike, like a mescaline vision—and, at the other extreme, the strange "organic" forms of kidney ore, hematite, looking so much like the kidneys of giant animals that I would wonder for a moment which museum I was in.

But finally I would always go back to the Science Museum, for this was the first one I had ever been to. My mother had sometimes brought my brothers and me here even before the war, when I was a child. She would lead us through the magical galleries—the early airplanes, the dinosaur-like machines of the Industrial Revolution, the old optical contrivances—to a smaller gallery at the top where there was a reconstruction of a coal mine with the original equipment. "Look!" she would say. "Look there!" And she'd direct our gaze to an old mining lamp. "My father, your grandfather, invented that!" she would say, and we would bend our heads and read: "The Landau lamp, invented by Marcus Landau in 1869. It displaced the earlier Humphry Davy lamp." Whenever I read this, it excited me strangely and gave me a sense of a personal bond to the museum and to my grandfather (born in 1837 and long since dead), the sense that he and his invention were still somehow real and alive.

But the real epiphany came for me in the Science Museum when I was ten, and I discovered the periodic table up on the fifth floor—not one of your nasty, natty, modern little spirals, but a solid rectangular one covering a whole wall, with separate cubicles for every element and the actual elements, whenever possible, in place: chlorine, greenish yellow; swirling brown bromine; jet-black (but violet-vapored) crystals of iodine; heavy, heavy slugs of uranium; and pellets of lithium floating in oil. They even had the inert gases (or "noble" gases, too noble to combine):

helium, neon, argon, krypton, xenon (but not radon—I guessed it was too dangerous). They were invisible, of course, inside their sealed glass tubes, but one knew they were there.

The actual presence of the elements reinforced the feeling that these were indeed the elemental building blocks of the universe, that the whole universe was here, in microcosm, in South Kensington. I had an overwhelming sense of Truth and Beauty when I saw the periodic table, a sense that this was not a mere human construct, arbitrary, but an actual vision of the eternal cosmic order, and that any future discoveries and advances, whatever they might add, would only reinforce, reaffirm, the truth of its order.

This feeling of grandeur, the immutability of nature's laws, and of how they might prove graspable by us if we sufficiently sought them—this came to me overwhelmingly when I was a boy of ten, standing before the periodic table in the Science Museum in South Kensington. It has never left me, and fifty years later it is undimmed. My faith and life were set at that moment; my Pisgah, my Sinai, came in a museum.

# First Love

In January 1946, when I was twelve and a half, I moved from my prep school in Hampstead, The Hall, to a much larger school, St. Paul's, in Hammersmith. It was here, in the Walker Library, that I met Jonathan Miller for the first time. I was hidden in a corner, reading a nineteenth-century book on electrostatics—reading, for some reason, about "electric eggs"—when a shadow fell across the page. I looked up and saw an astonishingly tall, gangling boy with a very mobile face, brilliant, impish eyes, and an exuberant mop of reddish hair. We got talking together, and have been close friends ever since.

Prior to this time, I had had only one real friend, Eric Korn, whom I had known almost from birth. Eric followed me from The Hall to St. Paul's a year later, and now he and Jonathan and I formed an inseparable trio, bound not only by personal but by family bonds, too (our fathers, thirty years earlier, had all been medical students together, and our families had remained close). Jonathan and Eric did not really share my love of chemistry—though a year or two earlier they had joined me in a flamboyant chemical experiment: throwing a large lump of metallic sodium into the Highgate Ponds on Hampstead Heath and watching excitedly as it took fire and sped round and round on the surface like a demented meteor, with a huge sheet of yellow flame beneath it—but they were intensely interested in biology, and it was inevitable, when the time came, that we would find ourselves together in the same biology class, and that all of us would fall in love with our biology teacher, Sid Pask.

Pask was a splendid teacher. He was also narrow-minded, bigoted, cursed with a hideous stutter (which we would imitate endlessly), and by no means exceptionally intelligent. By dissuasion, irony, ridicule, or force, he would turn us away from all other activities—from sport and sex, from religion and families, and from all our other subjects at school. He demanded that we be as single-minded as he was.

The majority of his pupils found him an impossibly demanding and exacting taskmaster. They would do all they could to escape from this pedant's petty tyranny, as they regarded it. The struggle would go on for a while, and then suddenly there was no longer any resistance—they were free. Pask no longer carped at them, no longer made ridiculous demands upon their time and energy.

Yet some of us, each year, responded to Pask's challenge. In return, he gave us all of himself—all his time, all his dedication, for biology. We would stay late in the evening with him in the Natural History Museum. We would sacrifice every weekend to plant-collecting expeditions. We would get up at dawn on freezing winter days to go on his January freshwater course. And once a year—there is still an almost intolerable sweetness about the memory—we would go with him to Millport for three weeks of marine biology.

Millport, off the western coast of Scotland, had a beautifully equipped marine biology station, where we were always given a friendly welcome and inducted into whatever experiments were going on. (Fundamental observations were being made on the development of sea urchins at this time, and Lord Rothschild, now in the midst of his soon-to-be-famous experiments on the fertilization of sea urchins, was endlessly patient with the enthusiastic schoolboys who crowded around and peered into his petri dishes with the transparent pluteus larvae.) Jonathan, Eric, and I made several transects on the rocky shore together, counting all the animals and seaweeds we could on successive square-foot portions, from the lichen-covered summit of the rock (*Xanthoria parietina* was the euphonious name of this lichen) to the shoreline and tidal pools below. Eric was particularly and wittily ingenious, and once, when we needed a plumb line to give us a true vertical but did not know how to suspend it, he pried a limpet from the base of a rock, placed the tip of the plumb line beneath it, and firmly reattached it at the top as a natural drawing pin.

We all adopted particular zoological groups: Eric became enamored of sea cucumbers, holothurians; Jonathan of iridescent bristled worms, polychaetes; and I of squids and cuttlefish, octopuses, all cephalopods—the most intelligent and, to my eyes, the most beautiful of invertebrates. One day we all went down to the seashore, to Hythe in Kent, where Jonathan's parents had taken a house for the summer, and went out for a day's fishing on a commercial trawler. The fishermen would usually throw back the cuttlefish that ended up in their nets (they were not popular eating in England). But I, fanatically, insisted that they keep them for me, and there must have been dozens of them on the deck by the time we came in. We brought all the cuttlefish back to the house in pails and tubs, put them in large jars in the basement, and added a little alcohol to preserve them. Jonathan's parents were away, so we did not hesitate. We would be able to take all the cuttlefish back to school, to Pask—we imagined his astonished smile as we brought them in—and there would be a cuttlefish apiece for everyone in the class to dissect, two or three apiece for the cephalopod enthusiasts. I myself would give a little talk about them at the Field Club, dilating on their intelligence, their large brains, their eyes with erect retinas, their rapidly changing colors.

A few days later, the day Jonathan's parents were due to return, we heard dull thuds emanating from the basement, and going down to investigate, we encountered a grotesque scene: the cuttlefish, insufficiently preserved, had putrefied and fermented, and the gases produced had exploded the jars and blown great lumps of cuttlefish all over the walls and floor; there were even shreds of cuttlefish stuck to the ceiling. The intense smell of putrefaction was awful beyond imagination. We did our best to scrape off the walls and remove the exploded, impacted lumps of cuttlefish. We hosed down the basement, gagging, but the stench was not to be removed, and when we opened windows and doors to air out the basement, it extended outside the house as a sort of miasma for fifty yards in every direction.

Eric, always ingenious, suggested we mask the smell, or replace it, with an even stronger but pleasant smell—a coconut essence, we decided, would fill the bill. We pooled our resources and bought a large bottle of this, which we used to douche the basement, then distributed liberally through the rest of the house and its grounds.

Jonathan's parents arrived an hour later and, advancing towards the house, encountered an overwhelming scent of coconut. But as they drew nearer they hit a zone dominated by the stench of putrefied cuttlefish—the two smells, the two vapors, for some curious reason, had organized themselves in alternating zones about five or six feet wide. By the time they reached the scene of our accident, our crime, the basement, the smell was insupportable for more than a few seconds. The three of us were all in deep disgrace over the incident. I especially, since it had arisen from my greed in the first place (would not a single cuttlefish have done?) and my folly in not realizing how much alcohol so many specimens would need. Jonathan's parents had to cut short their holiday and leave the house (the house itself, we heard, remained uninhabitable for months). But my love of cuttlefish remained unimpaired.

Perhaps there was a chemical reason for this, as well as a biological one, for cuttlefish (like many other mollusks and crustaceans) have blue blood, not red, because they evolved a completely different system for transporting oxygen from the one we vertebrates did. Whereas our red respiratory pigment, hemoglobin, contains iron, their bluish-green pigment, hemocyanin, contains copper. Iron and copper each have two different "oxidation states," and this means that they can easily take up oxygen in the lungs, move it to a higher oxidation state, and then relinquish it, in the tissues, as needed. But why employ just iron and copper when there was another metal—vanadium, a neighbor of theirs in the periodic table—that had no less than four oxidation states? I wondered if vanadium compounds were ever exploited as respiratory pigments, and got most excited when I heard that some sea squirts, tunicates, were extremely rich in the element vanadium and had special cells, vanadocytes, devoted to storing it. Why they contained these was a mystery; they did not seem to be part of an oxygen-transport system.

Absurdly, impudently, I thought I might solve this mystery during one of our annual excursions to Millport. But I got no further than collecting a bushel of sea squirts (with the same greed, the same inordinacy, that had caused me to collect too many cuttlefish). I could incinerate these, I thought, and measure the vanadium content of their ash (I had read that this could exceed 40 percent in some species). And this gave me the only commercial idea I have ever had: to open a vanadium farm—acres of sea meadows, seeded with sea squirts. I would get them to extract the precious vanadium from seawater, as they had been doing very efficiently for the last three hundred million years, and then sell it for £500 a ton. The only problem, I realized, aghast at my own genocidal thoughts, would be the veritable holocaust of sea squirts required.

# Humphry Davy:
# Poet of Chemistry

Humphry Davy was for me—as for most boys of my generation with a chemistry set or a lab—a beloved hero; a boy himself in the boyhood of chemistry; an intensely appealing figure, as fresh and alive after a hundred years in his way as anyone we knew. We knew all about his youthful experiments—from nitrous oxide (which he discovered, described, and became slightly addicted to as a teenager); to his often reckless experiments with alkali metals, electric batteries, electric fish, explosives. We imagined him as a Byronic young man with wide-set, dreaming eyes.

It happened that I was thinking of Humphry Davy when I saw a notice of David Knight's 1992 biography, *Humphry Davy: Science and Power,* and I immediately sent for it. I had been in a nostalgic mood, recalling my own boyhood: my twelve-year-old self most romantically and deeply in love—more deeply, perhaps, than ever again—with sodium and potassium and chlorine and bromine; in love with a magical shop in whose dark interior I could purchase chemicals for my lab; with the heavy, encyclopedic volume of Mellor (and where I could decipher them, the Gmelin handbooks); with London's Science Museum in South Kensington, where the history of chemistry, especially its beginnings in the late eighteenth and early nineteenth centuries, was laid out; in love, perhaps most of all, with the Royal Institution, much of which still looked and smelled exactly as it must have when the young Humphry Davy worked there, and where one could browse among and ponder his actual notebooks, manuscripts, lab notes, and letters.

Davy is, as Knight remarks, a wonderful subject for a biographer, and there have been many biographies of him in the last century and a half. But Knight—trained as a chemist, a professor of the history and philosophy of science at Durham, and former editor of the *British Journal for the History of Science,* has produced a work that is not only grand and scholarly but full of human insight and sympathy, too.

Davy was born in 1778 in Penzance, the eldest of five children, to an engraver and his wife. He went to the local grammar school and enjoyed its freedom. ("I consider it fortunate that I was left much to myself as a child, and put upon no particular plan of study," he noted.) He left school at sixteen and was apprenticed to a local apothecary-surgeon, but he was bored by this and aspired to something larger. Chemistry, above all, started to attract him: he read and mastered Lavoisier's great *Elements of Chemistry* (1789), a remarkable achievement for an eighteen-year-old with little formal education. Grand visions started revolving in his mind: Could he be the new Lavoisier, perhaps the new Newton? One of his notebooks from this time was labeled "Newton and Davy."

And yet, in a way, it was less with Newton than with Newton's friend and contemporary Robert Boyle that Davy's affinities lay. For while Newton had founded a new physics, Boyle had founded the equally new science of chemistry and disentangled it from its alchemical precursors. It was Boyle, in his 1661 *Sceptical Chymist,* who threw out the metaphysical four elements of the ancients and redefined "elements" as simple, pure, undecomposable bodies made up of "corpuscles" of a particular kind. It was Boyle who saw the main business of chemistry as analysis (and who introduced the word "analysis" in a chemical context), breaking down complex substances into their constituent elements and seeing how these could combine. Boyle's enterprise gathered force in the late seventeenth and early eighteenth centuries, when more than a dozen new elements were isolated in quick succession.

But a peculiar confusion attended the isolation of these elements. The Swedish chemist Carl Wilhelm Scheele obtained a heavy greenish vapor from hydrochloric acid in 1774 but failed to realize that it was an element. He saw it instead as "dephlogisticated muriatic acid." Joseph Priestley, isolating oxygen the same year, called that gas "dephlogisticated air." These misinterpretations arose from a half-mystical theory that had dominated chemistry throughout the eighteenth century and, in many ways, prevented its advance. "Phlogiston" was, it was believed, an immaterial substance given off by burning bodies; it was the material of heat.

Lavoisier, whose *Elements* was published when Davy was eleven, overthrew the phlogiston theory and showed that combustion did not involve the loss of a mysterious "phlogiston" but resulted instead from the combination of what was burned with oxygen from the atmosphere (or oxidation).

Lavoisier's work stimulated Davy's first, seminal experiment at the age of eighteen, when he melted ice by friction, thus showing that heat was energy, and not a material substance like caloric. "The non-existence of caloric, or the fluid of heat, has been proved," he exulted. Davy embodied the results of his experiments in a long work titled "An Essay on Heat, Light, and the Combinations of Light," which included a critique of Lavoisier and of all chemistry since Boyle, as well as a vision of a new chemistry that he hoped to found, one purged of all the metaphysics and phantoms of the old.

News of the young man and his revolutionary new thoughts about matter and energy, reached Thomas Beddoes, then a professor of chemistry at Oxford. Beddoes invited Davy to his laboratory in Bristol, and here Davy did his first major work, isolating the oxides of nitrogen and examining their physiological effects.[1]

Davy's period at Bristol saw the start of his close friendship with Coleridge and the Romantic poets. He was writing a good deal of poetry himself at the time, and his notebooks mix details

---

1. This included a wonderful account of the effects of inhaling the fumes of nitrous oxide—"laughing gas"—which in its psychological perspicacity is reminiscent of William James's own account of the same experience, a century later. It is perhaps the first description of a psychedelic experience in Western literature:

> A thrilling extending from the chest to the extremities was almost immediately produced . . . my visible impressions were dazzling and apparently magnified, I heard distinctly every sound in the room. . . . As the pleasurable sensations increased, I lost all connection with external things; trains of vivid visible images rapidly passed through my mind and were connected with words in such a manner, as to produce perceptions perfectly novel. I existed in a world of newly connected and newly modified ideas. I theorised; I imagined that I made discoveries.

Davy also discovered that nitrous oxide was an anesthetic and suggested its use in surgical operations. He never followed up on this, and general anesthesia was introduced only in the 1840s, after his death. (Freud, in the 1880s, was similarly careless of his own discovery that cocaine was a local anesthetic—and the credit for this discovery is usually given to others.)

of chemical experiments, poems, and philosophical reflections all together. Joseph Cottle, who had published Coleridge and Southey, felt that Davy was a poet no less than a natural philosopher, and that either, or both, represented his singularity of perception: "It was impossible to doubt, that if he had not shone as a philosopher, he would have become conspicuous as a poet." Indeed, in 1800, Wordsworth asked Davy to oversee the publication of the second edition of his *Lyrical Ballads*.

At this time there still existed a union of literary and scientific cultures; there was not the dissociation of sensibility that was so soon to come. There was indeed, between Coleridge and Davy, a close friendship and a sense of almost mystical affinity and rapport. The analogy of chemical transformation leading to the emergence of wholly new compounds was central to Coleridge's thinking, and at one point he planned to set up a chemical laboratory with Davy. The poet and the chemist were fellow warriors, analyzers and explorers of a principle of connectedness of mind and nature.[2]

Coleridge and Davy seemed to see themselves as twins: Coleridge the chemist of language, Davy the poet of chemistry.

CHEMISTRY WAS CONCEIVED, in Davy's time, to embrace not only chemical reactions proper but the study of heat, light, mag-

---

2. In Coleridge's words,

> Water and flame, the diamond, the charcoal . . . are convoked and fraternized by the theory of the chemist. . . . It is the sense of a principle of connection given by the mind, and sanctioned by the correspondency of nature. . . . If in a *Shakespeare* we find nature idealized into poetry . . . so through the meditative observation of a *Davy* . . . we find poetry, as it were, substantiated and realized in nature: yea, nature itself disclosed to us, . . . as at once the poet and the poem!

netism, and electricity—much of what was later to be separated off as "physics." (Even at the end of the nineteenth century, the Curies first regarded radioactivity as a "chemical" property of certain elements.) And though static electricity was known in the eighteenth century, no sustained electric current was possible until Alessandro Volta invented a sandwich of two different metals with brine-dampened cardboard in between, which generated a steady electric current—the first battery. Davy later wrote that Volta's paper, published in 1800, acted like an alarm bell among the experimenters of Europe, and for Davy, it suddenly gave form to what he would now see as his life's work.

He persuaded Beddoes to build a large electric battery modeled after Volta's, and started his first experiments with it in 1800. He suspected almost at once that its current was generated by chemical changes in the metal plates and wondered if the reverse was also true: could one induce chemical changes by the passage of an electric current? He made ingenious and radical modifications to the battery, and he was the first to make use of the enormous new power available to devise a new form of illumination, the carbon arc lamp.

These brilliant advances excited attention in the capital, and in that same year Davy was invited to the newly founded Royal Institution in London. He had always been eloquent and a natural storyteller, and now he was to become the most famous and influential lecturer in England, drawing huge crowds that blocked the streets whenever he lectured. His lectures moved from the most intimate details of his experiments—reading them gives a vivid view of the work in progress, of the activity of an extraordinary mind—to speculation about the universe and life, delivered in a style and with a richness of language that nobody else could match.

Davy's inaugural lecture enthralled many, including Mary Shelley. Years later, in *Frankenstein,* she was to model Professor Waldman's lecture on chemistry rather closely on some of Davy's words. (Specifically, when, speaking of galvanic electricity, Davy had said, "A new influence has been discovered, which has enabled man to produce from combinations of dead matter effects which were formerly occasioned only by animal organs.") And Coleridge, the greatest talker of his age, always came to Davy's lectures, not only to fill his chemical notebooks but, as he said, "to renew my stock of metaphors."[3]

There was an extraordinary appetite for science, especially chemistry, in the early, palmy days of the Industrial Revolution; it seemed a new and powerful (and not irreverent) way not only of understanding the world but moving it to a better state. This double view of science found its perfect exponent in Davy.

IN THESE FIRST YEARS of the Royal Institution, Davy put aside his larger speculations and concentrated on particular practical problems: problems of tanning and the isolation of tannin (he was the first to find it in tea) and a whole range of agricultural problems—he was the first to recognize the vital role of nitrogen

3. Coleridge was not the only poet to renew his stock of metaphors with images from chemistry. The chemical phrase "elective affinities" was given an erotic connotation by Goethe; "energy" became, for Blake, "eternal delight"; Keats, trained in medicine, also reveled in chemical metaphors.
    Eliot, in "Tradition and the Individual Talent," employs chemical metaphors from beginning to end, culminating in a grand, Davyan metaphor for the poet's mind: "The analogy is that of the catalyst. . . . The mind of the poet is the shred of platinum." One wonders whether Eliot knew that his central metaphor, catalysis, was discovered by Humphry Davy in 1816.

and the importance of ammonia in fertilizers (his *Elements of Agricultural Chemistry* was published in 1813).

By 1806, however, established as the most brilliant lecturer and practical chemist in England—and still only twenty-seven—Davy felt he needed to give up his research obligations at the Royal Institution and return to the fundamental concerns of his Bristol days. He had long wondered whether an electric current could provide a new way of isolating chemical elements, and he began experimenting with the electrolysis of water, using an electric current to split it into its component elements of hydrogen and oxygen and showing that these combined in exact proportions.

The following year he performed the famous experiments that isolated metallic potassium and sodium by electric current. When the current flowed, Davy wrote, "a most intense light was exhibited at the negative wire, and a column of flame . . . arose from the point of contact." This produced shining metallic globules, indistinguishable in appearance from mercury—globules of two new elements, potassium and sodium. "The globules often burnt at the moment of their formation," he observed, "and sometimes violently exploded and separated into smaller globules, which flew with great velocity through the air in a state of vivid combustion, producing a beautiful effect of continued jets of fire." When this occurred, Davy, his cousin Edmund records, danced with joy around the lab.[4]

My own greatest delight as a boy was to repeat Davy's electrolytic production of sodium and potassium, to see these shin-

4. Davy was so startled by the inflammability of sodium and potassium, and their ability to float on water, that he wondered whether there might not be deposits of these beneath the earth's crust, which, exploding upon the impact of water, were responsible for volcanic eruptions.

ing globules catch fire in the air, burning with a vivid yellow flame or a pale mauve one, and later, to obtain metallic rubidium (which burns with an enchanting ruby-red flame)—an element not known to Davy, but one he would certainly have appreciated. I so strongly identified with Davy's original experiments that I could almost imagine I was discovering these elements myself.

Davy turned to the alkaline earths next, and within a few weeks had isolated their metallic elements, too—calcium, magnesium, strontium, and barium. These were highly reactive metals, especially strontium and barium, able to burn, like the alkali metals, with brilliantly colored flames. And if the isolation of six new elements in a single year was not enough, Davy isolated yet another element, boron, the following year.

ELEMENTAL SODIUM and potassium do not exist in nature; they are too reactive and will instantly combine with other elements. What one finds, instead, are salts—sodium chloride (common salt), for example—compounds that are chemically inert and electrically neutral. But if one submits these, as Davy did, to a powerful electric current transmitted through two electrodes, the neutral salt can be decomposed as its electrically charged particles (electropositive sodium, electronegative chloride, in this case) are attracted towards either electrode. (Faraday later named these particles "ions.")

For Davy, electrolysis was not only "a new path to discovery" that incited him to request ever larger and more powerful batteries for his use. It was also a revelation that matter itself was not something inert, as Newton and others had thought, but was charged and held together by electrical forces.

Chemical affinity and electrical force, Davy now realized, determined each other, and were one and the same in the constitution of matter. Boyle and his successors, including Lavoisier, had no clear idea about the fundamental nature of chemical bonds, but they were assumed to be gravitational. Davy could now envisage another universal force, electrical in nature, holding together the very molecules of matter itself. Beyond this, he had a cloudy but intense vision that the entire cosmos was pervaded by electrical forces as well as gravitation.

In 1810, Davy reexamined Scheele's heavy greenish gas, previously seen by Scheele and Lavoisier as compound in nature, and he was able to show that it was an element. He named it chlorine, in view of its color (from the Greek *chloros,* greenish yellow). He realized that it was not only a new element but a representative of a whole new chemical family—a family of elements like the alkali metals, too active to exist in nature. Davy felt sure there must be heavier and lighter analogues of chlorine, members of the same family.

THESE YEARS FROM 1806 to 1810 were the most creative years of Davy's life, both in his empirical discoveries and in the profound concepts arising from them. He had discovered eight new elements. He had overturned the last traces of the phlogiston theory and Lavoisier's notion that atoms were merely metaphysical entities. He had shown the electrical basis of chemical reactivity. He had grounded chemistry and transformed it, in these five intense years.

If he enjoyed the highest esteem from his colleagues, winning many scientific honors, he enjoyed an equal fame with the educated public through his popularizations of science. He loved to conduct experiments in public, and his famous lecture-demonstrations were exciting, eloquent, highly dramatic, and sometimes literally explosive. Davy seemed to be at the crest of a vast new wave of scientific and technological power, a power that promised, or threatened, to transform the world. What honor could the nation bestow on such a man? There seemed only one, though it was almost without precedent. On April 8, 1812, Davy was knighted by the prince regent, the first scientist to be so elevated since Newton in 1705.[5]

DAVY "CONDUCTED HIS RESEARCH in romantic disorder," Knight tells us, "and in great bursts of speed after an incubation period." He worked alone, aided only by a laboratory assistant. The first of these was his younger cousin Edmund Davy; the second was Michael Faraday, whose relationship to Davy was to become an intense and complex one, passionately positive at first, clouded later. Faraday was almost a son to Humphry Davy, "a son in science," as the French chemist Berthollet was to say of his own "son," Gay-Lussac. Faraday, then in his early twenties, had followed Davy's lectures raptly, and wooed Davy by presenting him with a brilliantly transcribed and annotated version of them.

Davy hesitated before taking Faraday on as his assistant. Faraday was an unknown quantity; he was shy, unworldly, gauche, poorly educated. But he had an intense, precocious love of science and an extraordinary brain. He was in many ways

5. The term "scientist" did not exist in 1812. The great historian of science William Whewell devised it in 1834.

like Davy himself when he had approached Beddoes. Davy was initially a generous and supportive "father" but later, with Faraday's increasing intellectual independence, became an oppressive and perhaps envious one.

Faraday, at first wholly admiring of the older man, grew increasingly resentful and also felt a moralistic contempt for Davy's worldliness. An adherent of a fundamentalist religious sect, he disapproved of all titles, honors, and offices, and resolutely refused them himself in later life. And yet at a deeper level there was between the two men an affection and an intellectual intimacy that never fully deserted them. Both men being shy and somewhat formal in utterance, it is impossible to do more than guess at the inner history of their relationship. But the creative encounter between these two minds of the highest caliber in a sustained and intense relationship was of the greatest importance to both and, indeed, to the history of science.

DAVY HAD STRONG AMBITIONS for social status and prestige and power, and three days after he was knighted, he married Jane Apreece, a well-connected, bluestocking heiress and a cousin of Sir Walter Scott. Lady Davy (as Sir Humphry always referred to her) was a brilliantly articulate woman who had had a salon in Edinburgh, but like Davy, she was used to independence and adulation; neither was suited to domestic life. The marriage was not only unhappy but destructive of Davy's dedication to science. More and more of his energy was devoted to hobnobbing with and emulating the aristocrats ("he dearly loved a Lord," Knight remarks) and trying to be one himself—a hopeless task in Regency England, where a man's class was ineluctably ordained by his birth, and neither eminence nor title nor marriage could change this.

The Davys did not immediately go on their honeymoon but planned instead to spend a year on the Continent together as soon as Humphry had completed his current researches. He had been working on gunpowder and other explosives, and in October of 1812 he experimented with the first "high" explosive, nitrogen trichloride, which has cost many people fingers and eyes. He discovered several new ways of making the combination of nitrogen and chlorine, and caused a violent explosion on one occasion while he was visiting a friend. He wrote all the details to his admiring brother, John: "It must be used with very great caution. It is not safe to experiment upon a globule larger than a pin's head. I have been severely wounded by a piece scarcely bigger."

Davy himself was partially blinded and did not recover fully for another four months. We are not told what damage was done to his friend's house.

The honeymoon was bizarre and comic at the same time. Davy brought along a good deal of chemical apparatus and various materials: "an air pump, an electrical machine, a voltaic battery . . . a blow-pipe apparatus, a bellows and forge, a mercurial and water gas apparatus, cups and basins of platinum and glass, and the common reagents of chemistry," to which he added some high explosives to experiment with. He also brought along his young research assistant, Faraday (who was treated like a servant by Lady Davy and soon came to hate her).

In Paris, Davy had a visit from Ampère and Gay-Lussac, who brought with them, for his opinion, a sample of a shiny black substance with the remarkable property that when heated, it did not melt, but turned at once into a vapor of a deep violet color. Davy sensed that this might be an analogue of chlorine and soon confirmed that it was a new element ("a new species of matter," as he wrote in his report to the Royal Society), to which he gave another chromatic name: iodine, from the Greek *ioeides,* violet-colored.

From France the wedding party moved by stages to Italy, with experiments along the way: burning a diamond, under controlled conditions, with a giant magnifying glass in Florence;[6] collecting crystals from the rim of Vesuvius; analyzing gas from natural vents in the mountains—it turned out to be, Davy found, identical with marsh gas, or methane; and, for the first time, analyzing samples of paint from old masterworks ("mere atoms," Davy announced).

During this strange chemical honeymoon-à-trois, traipsing across Europe, Davy seemed to revert to an irrepressible, inquisitive, mischievous boy full of ideas and pranks. It was a wonderful induction into the scientific life for Faraday, though Lady Davy, it seems, was indisposed for much of the time. But the holiday, long extended, had to come to an end, and the titled couple returned to London, where Davy took on the grandest practical challenge of his lifetime.

The Industrial Revolution, now warming up, devoured ever huger amounts of coal; coal mines were dug deeper, deep enough to run into the inflammable and poisonous gases of "fire-damp" (methane) and "choke-damp" (carbon dioxide). A canary car-

6. Davy had been reluctant, up to this point, to believe that diamond and charcoal were, in fact, one and the same element; he felt this was "against the analogies of Nature." It was perhaps his weakness, as well as his strength, that he sometimes thought to classify the chemical world by concrete qualities, not formal properties. (For the most part—as with the alkali metals and the halogens—concrete qualities correspond to formal properties; it is rather rare for elements to have a number of quite different physical forms.)

ried down in a cage could serve as a warning of the presence of asphyxiating choke-damp, but the first indication of fire-damp was all too often a fatal explosion. It was desperately important to design a miner's lamp that could be carried into the lightless depths of the mines without any danger of igniting pockets of fire-damp.

Davy experimented with many different designs for his lamp, and in so doing discovered a number of new principles. He found that the use of narrow metal tubes, in airtight lanterns, prevented the propagation of explosions. He then experimented with wire gauzes, and found that flames could not pass these.[7] Using tubes and gauzes, the perfected Davy lamps, tested in 1816, not only proved safe but also, by the appearance of the flame, were reliable indicators of the presence of fire-damp.[8]

Davy never sought compensation or patented his invention of the safety lamp, but gave it freely to the world. (In this he was a contrast to his friend William Hyde Wollaston, who made a huge fortune through his commercial exploitation of palladium and platinum.)

This was the high point of Davy's public life, as his electrochemical researches had been the high point of his intellectual

7. Davy went on with his investigations of flame and, a year after the safety lamp, published "Some New Researches on Flame." More than forty years later, Faraday would return to the subject, in his famous 1861 Royal Institution lecture series on *The Chemical History of a Candle.*

8. This was my own introduction to Humphry Davy as a child, when my mother took me to the Science Museum in London, up to the top floor where there was a very realistic simulacrum of a nineteenth-century coal mine. She showed me the Davy lamp, and explained how it made it safer to work in coal mines; then she showed me another safety lamp, the Landau lamp. "My father, your grandfather, invented this," she said, "when he was a young man in 1869. It was even safer than the original design, and came to replace the Davy lamp." I felt a thrill of identification. I had then the sense—childish, but very vivid—of science as a completely human business: influences, conversations, across the ages.

life. With the creation of his safety lamp and its gift to the nation, public awareness and approbation rose to new heights.

THERE WAS A VISIONARY, mystical dimension to Davy, not evident to his contemporaries (save perhaps Coleridge and Faraday, who knew him so well, and who were so great and so strange in their own ways), hidden behind the dazzle of his practical achievements.

Davy took great pains to be an empiricist, but he was also a part of the Romantic movement and its *Naturphilosophie* and remained so throughout his life. There is not necessarily any contradiction between a mystical or transcendent philosophy and a rigorously empirical mode of experiment and observation; they can go together, as they certainly did with Newton. Davy had been fascinated by idealistic philosophy as a young man, benefiting from Coleridge's passionate translations of Friedrich Schelling, and his own work served to provide an empirical confirmation of some of Schelling's notions: that the universe was a dynamic whole, held together by energies of opposite valence, and one in which energy, however transformed, was always conserved.

For Newton, space was a mere medium, structureless, in which motion occurred, while forces such as gravity were quite mysterious, seeming to exemplify "action at a distance." Only with Faraday came the notion that forces have structure, that magnets or current-bearing wires create a charged field. But it seems to me that Davy was close to the concept of "field"—the transcendent and, in a sense, Romantic concept we owe to Faraday. One wonders what passed between these two visionary geniuses, Faraday and Davy, when—greatly excited by the work of Ørsted, Ampère, and others—they thought together on the newly discovered phenomena of electromagnetism. It is tantalizing to think of Davy as a junctional figure between the idealistic universes of Leibniz and Schelling and the modern universes of Faraday, Clerk Maxwell, and Einstein.

IN 1820, Davy was accorded the highest honor in science: the presidency of the Royal Society. Newton had held this position for twenty-four years; and the incumbent before Davy, for forty-two years, had been the aristocratic Sir Joseph Banks. No office in science carried more power or prestige, but none carried heavier diplomatic or administrative burdens. It has been estimated that Banks wrote more than fifty thousand letters, and perhaps as many as a hundred thousand, during his tenure. This crushing burden now fell on Davy.

Even more serious were the repercussions of Davy's efforts to reform the Royal Society, which, by the 1820s, had to some extent become a society of well-born, sometimes highly gifted men who had not actually done anything much for science. Davy argued, not too tactfully, that the society had been losing its reputation steadily and that its fellows must prove their worth. His constant, often uncouth efforts to diminish unproductive patronage and to shape a society of amateurs and gentlemen into professionals caused defiance and anger among many of the fellows. Davy increasingly became the object of scorn and hostility, and he who had once been described as "enchanting" in manner reacted to all this with rage, arrogance, and intransigence. One sees the bloated, red-faced rage in the portrait of him from this time that hangs in the Royal Institution. Once the most popular scientist in England, he became, in David Knight's words, "one of the most disliked men of science ever."

These were evil times for Davy. Continually vexed with the trivia of the Royal Society; at bay with most of its fellows; cut off from Coleridge and other friends with whom in earlier days he had known such openness and happiness; stuck in a loveless, childless marriage; conscious, increasingly, as he moved through his forties, of vague organic symptoms, intimations perhaps of the problems which had brought his father to an early death, Davy had reason to bewail his state and to look back to the powers of an earlier time. He was too distracted to do any original work, which had always been his chief and often only source of inner peace and stability; worse, he no longer felt himself in the forefront of his subject, perceiving that he was regarded by his contemporaries as obsolete or marginal. The Swedish chemist Berzelius, who was now bringing all of inorganic chemistry under his sway, dismissed Davy's life work as no more than "brilliant fragments."

His sense of loss, of hopeless nostalgia, deepened each year. "Ah!" he wrote in 1828,

> Could I recover anything like that freshness of mind, which I possessed at twenty-five . . . what would I not give! . . . How well I remember that delightful season, when, full of power, I sought for power in others; and power was sympathy, and sympathy power;—when the dead and the unknown, the great of other ages and distant places, were made, by the force of the imagination, my companions and friends.

. . .

IN 1826, Davy's mother died. He was singularly attached to her, as Newton had been to his mother, and her death affected him grievously. Later that year, at the age of forty-eight, he suffered, as his father had at the same age, a transient numbness in his hand and arm, and weakness in his leg, followed by a paralytic stroke. Though he recovered speedily, the gravity of this, and its undeniable import, altered his thinking. He suddenly felt sick of the endless struggles at the Royal Society, the endless obligations of his worldly life: "My health was gone, my ambition satisfied, I was no longer excited by the desire of distinction; what I regarded most tenderly was in the grave."

One of Davy's recreations, perhaps his only one, throughout his adult life, had been fishing. Otherwise distracted, or pompous, or unapproachable, he would regain all his old friendliness, his real self, when fishing. This was the time when his mind became youthful and fresh once again, and he could delight, as he used to, in the pure play of ideas. Over the years Davy, an expert fisherman, became equally expert in his knowledge of flies and fishes. One of his final, meditative books, *Salmonia,* is at once a natural history, an allegory, a dialogue, a poem; Knight calls it "a fishing book suffused with natural theology."

After completing the book, Davy set sail for Slovenia, accompanied by his godson John Tobin, the last of his scientific "sons." Out of England and its climate, which, Davy felt, kept "the nervous system in a constant state of disturbance," he might hope to receive, to enjoy, and to communicate his final thoughts: "I had sought for and found consolation, and partly recovered my health after a dangerous illness . . . I had found the spirit of my early vision. . . . Nature never deceives us; the rocks, the mountains, the streams, always speak the same language."

After his final, mortal stroke, in February 1829, he dictated this letter, his *Nunc Dimittis:*

> I am dying from a severe attack of palsy, which has seized the whole of the body, with the exception of the intellectual organ. . . . I bless God that I have been able to finish my intellectual labours.

I HAVE SAID THAT Humphry Davy was a boyhood hero to virtually everyone interested in chemistry or science in my generation. We all knew and repeated his famous experiments, imagining ourselves in his place. Davy himself had had such ideal companions in his youth, particularly Newton and Lavoisier. Newton, for him, was a sort of god; but Lavoisier was closer, more like a father with whom he could talk, agree, disagree. His own first essay, which Beddoes had published, while taking strong issue with Lavoisier, was in effect a dialogue with him. All of us need such figures, such ego ideals, and need them throughout life.[9]

Now I find to my dismay that when I speak to my younger scientific friends, none of them has heard of Davy, and some of them are puzzled when I tell them of my interest. It is difficult for them to imagine what relevance such "old" science can have. Science, it is often said, is impersonal, consists of "information" and "concepts"; these advance by a process of revision and replacement in which old information and old concepts become

9. The general theme of ego ideals, and the universal need for them, is especially explored in the opening chapter ("Making Great Men Ours") of Leonard Shengold's book *The Boy Will Come to Nothing! Freud's Ego Ideal and Freud as Ego Ideal.*

obsolete. In this view, the science of the past is irrelevant to the present, of interest only to the historian or psychologist.

But this is not what I have found in reality: when I came to write my first book, *Migraine,* in 1967, I was stimulated by the nature of the malady and by encounters with my patients, but equally, and crucially, by an "old" book on the subject, Edward Liveing's *Megrim,* written in the 1870s. I took this book out of the rarely entered historical section of the medical school library and read it, cover to cover, in a sort of rapture. I reread it many times for six months, and I got to know Liveing extremely well. His presence and his way of thinking were continually with me. My prolonged encounter with Liveing was crucial for the generation of my own thoughts and book. It was just such an encounter with Humphry Davy, when I was twelve, that had originally confirmed me on the path to science. How could I believe that the history of science, the past, was irrelevant?

I do not think my experience is unique. Many scientists, no less than poets or artists, have a living relation to the past, not just an abstract sense of history and tradition but a feeling of companions and predecessors, ancestors with whom they enjoy a sort of implicit dialogue. Science sometimes sees itself as impersonal, as "pure thought," independent of its historical and human origins. It is often taught as if this were the case. But science is a human enterprise through and through, an organic, evolving, human growth, with sudden spurts and arrests, and strange deviations, too. It grows out of its past but never outgrows it, any more than we outgrow our childhoods.

# Libraries

When I was a child, my favorite place at home was the library, a large oak-paneled room with all four walls covered by bookcases—and a solid table for writing and studying in the middle. It was here that my father had his special collection of books, as a Hebrew scholar; here, too, were all of Ibsen's plays (my parents had originally met in a medical students' Ibsen society); here, on a single shelf, were the young poets of my father's generation, many killed in the Great War; and here, on the lower shelves so I could easily reach them, were the adventure and history books belonging to my three older brothers. It was here that I found Kipling's *Jungle Book;* I identified deeply with Mowgli and used his adventures as a taking-off point for my own fantasies.

My mother had her favorite volumes in a separate bookcase in the lounge—Dickens, Trollope, and Thackeray, Bernard Shaw's plays in pale green bindings, as well as an entire set of Kipling bound in soft red morocco. There was a beautiful three-volume set of Shakespeare's works, a gilt-edged Milton, and other books, mostly poetry, that my mother had got as school prizes.

Medical books were kept in a special locked cabinet in my

parents' consulting room (but the key was in the door, so it was easy to unlock).

The oak-paneled library was the quietest and most beautiful room in the house, to my eyes, and it vied with my little chemistry lab as my favorite place to be. I would curl up in a chair and become so absorbed in what I was reading that all sense of time would be lost. Whenever I was late for lunch or dinner I could be found, completely enthralled by a book, in the library. I learned to read early, at three or four, and books, and our library, are among my first memories.

But the ur-library, for me, was our local public library, the Willesden library. There I spent many of the happiest hours of my growing-up years—our house was a five-minute walk from the library—and it was there I received my real education.

On the whole, I disliked school, sitting in class, receiving instruction; information seemed to go in one ear and out the other. I could not be passive—I had to be active, learn for myself, learn what *I* wanted, and in the way that suited me best. I was not a good pupil, but I was a good learner, and in the Willesden library—and all the libraries that came later—I roamed the shelves and stacks, had the freedom to select whatever I wanted, to follow paths that fascinated me, to become myself. At the library I felt free—free to look at the thousands, tens of thousands, of books; free to roam and to enjoy the special atmosphere and the quiet companionship of other readers, all, like myself, on quests of their own.

As I got older, my reading was increasingly biased towards the sciences, especially astronomy and chemistry. St. Paul's School, where I went when I was twelve, had an excellent general library, the Walker Library, which was particularly heavy in history and politics—but it could not provide all of the science and, especially, chemistry books I now hungered for. But with a special testimonial from one of the school masters, I was able to get a ticket to the library of the Science Museum, and there I devoured the many volumes of Mellor's *Comprehensive Treatise on Inorganic and Theoretical Chemistry* and the even longer *Gmelin Handbook of Inorganic Chemistry*.

When I went to university, I had access to Oxford's two great university libraries, the Radcliffe Science Library and the Bodleian, a wonderful general library that could trace itself back to 1602. It was in the Bodleian that I stumbled upon the now-obscure and forgotten works of Theodore Hook, a man greatly admired in the early nineteenth century for his wit and his prolific genius for theatrical and musical improvisation. I became so fascinated by Hook that I decided to write a sort of biography or "case history" of him. No other library—apart from the British Museum's library—could have provided the materials I needed, and the tranquil atmosphere of the Bodleian was a perfect one in which to write.

But the library I most loved at Oxford was our own library at the Queen's College. The magnificent library building itself had been designed by Christopher Wren, and beneath this, in an underground maze of heating pipes and shelves, were the vast subterranean holdings of the library. To hold ancient books, incunabula, in my own hands was a new experience for me—I particularly adored Gesner's *Historiae Animalium* (1551), richly illustrated with many wonderful engravings, including Dürer's drawing of a rhinoceros, and Agassiz's four-volume work on fossil fishes. It was there, too, that I saw all of Darwin's works in their original editions, and it was in the stacks that I found and fell in love with all the works of Sir Thomas Browne—his *Religio Medici,* his *Hydriotaphia,* and *The Garden of Cyrus (The Quincunciall Lozenge).* How absurd some of these were, but how magnificent the language! And if Browne's classical magniloquence became too much at times, one could switch to the lapidary cut and thrust of Swift—all of whose works, of course, were there in their original editions. While I had grown up on the nineteenth-century works my parents favored, it was the catacombs of the Queen's College library that introduced me to seventeenth- and eighteenth-century literature—Johnson, Hume, Pope, and Dryden. All of these books were freely available, not in some special, locked-away rare books enclave, but just sitting on the shelves, as they had done (I imagined) since their original publication. It was in the vaults of the Queen's College that I really gained a sense of history, and of my own language.

I first came to New York City in 1965, and at that time I had a horrid, poky little apartment in which there were almost no surfaces to read or write on. I was just able, holding an elbow awkwardly aloft, to write some of *Migraine* on the top of the refrigerator. I longed for spaciousness. Fortunately, the library at the Albert Einstein College of Medicine, where I worked, had this in abundance. I would sit at a large table to read or write for a while, and then wander around the shelves and stacks. I never knew what my eyes might alight upon, but I would sometimes discover unexpected treasures, lucky finds, and bring these back to my seat.

Though the library was quiet, whispered conversations might start in the stacks—two of you, perhaps, were searching for the same old book, the same bound volumes of *Brain* from 1890—and conversations could lead to friendships. All of us in the library were reading our own books, absorbed in our own worlds, and yet there was a sense of community, even intimacy. The physicality of books—along with their places and their neighbors on the bookshelves—was part of this camaraderie: handling books, sharing them, passing them to one another, even seeing the names of previous readers and the dates they took books out.

But a shift was occurring by the 1990s. I would continue to visit the library frequently, sitting at a table with a mountain of books in front of me, but students increasingly ignored the bookshelves, accessing what they needed with their computers. Few of them went to the shelves anymore. The books, so far as they were concerned, were unnecessary. And since the majority of users were no longer using the books themselves, the college decided, ultimately, to dispose of them.

I had no idea that this was happening—not only in the Einstein library but in college and public libraries all over the country. I was horrified when I visited the library recently and found the shelves, once overflowing, now sparsely occupied. Over the last few years, most of the books, it seems, have been thrown out, with remarkably little objection from anyone. I felt that a murder, a crime had been committed: the destruction of centuries of knowledge. Seeing my distress, a librarian reassured me that everything "of worth" had been digitized. But I do not use a computer, and I am deeply saddened by the loss of books, even bound periodicals, for there is something irreplaceable about a physical book: its look, its smell, its heft. I thought of how the library once cherished "old" books, had a special room for old and rare books; and how in 1967, rummaging through the stacks, I had found an 1873 book, Edward Liveing's *Megrim*, which inspired me to write my own first book.

# A Journey Inside the Brain

I first read Frigyes Karinthy's *A Journey Round My Skull* as a boy of thirteen or fourteen—I think it influenced me when I came to write my own neurological case histories—and now, rereading it sixty years later, I think it stands up remarkably well. It is not just an elaborate case history; it depicts the complex impact of a sight-, mind-, and life-threatening illness in a man of extraordinary sensibility and talent, and even something approaching genius, in the prime of his life. It becomes a journey of insight, of symbolic stages.

It has its faults: there are long digressions, philosophical and literary, where one might want a tauter narrative, and there is a certain amount of fanciful contrivance and extravagance—though this is something that Karinthy becomes more and more conscious of as he writes the book, as he is sobered by his experience, and as he tries to weld his novelistic imagination to the factual, even the clinical, realities of his situation. But despite its flaws, Karinthy's book is, to my mind, a masterpiece. We are inundated now with medical memoirs, both biographical and autobiographical—the entire genre has exploded in the last twenty years. Yet even though medical technology may have changed, the human experience has not, and *A Journey Round My Skull,* the first autobiographical description of a journey inside the brain, remains one of the very best.

FRIGYES KARINTHY, born in 1887, was a well-known Hungarian poet, playwright, novelist, and humorist when, at the age of forty-eight, he developed what in retrospect were the first symptoms of a growing brain tumor.

He was having tea at his favorite café in Budapest one evening when he heard "a distinct rumbling noise, followed by a slow, increasing reverberation . . . a louder and louder roar . . . only to fade gradually into silence." He looked up and was surprised to see that nothing was happening. There was no train; nor, indeed, was he near a train station. "What were they playing at?" Karinthy wondered. "Trains running outside . . . or some new means of locomotion?" It was only after the fourth "train" that he realized he was having a hallucination.

In his memoir Karinthy reflects on how he has occasionally heard his own name whispered softly—we have all had such experiences. But this was something quite different:

> The roaring of a train [was] loud, insistent, continuous. It was powerful enough to drown real sounds. . . . After a while I realized to my astonishment that the outer world was not responsible . . . the noise must be coming from inside my head.

Many patients have described to me how they first experienced auditory hallucinations—usually not voices or noises, but music. All of them, like Karinthy, looked around to find the source of what they were hearing, and only when they could find no possible source did they, reluctantly and sometimes fearfully, conclude that they were hallucinating. Many people in this situation fear that they are going insane—for is it not typical of madness to "hear things"?

Karinthy was not concerned on this score:

> I . . . did not find the incident at all alarming, but only very odd and unusual. . . . I could not have gone mad for, in that event, I should be incapable of diagnosing my case. Something else must be wrong.

SO THE FIRST CHAPTER of his memoir ("The Invisible Train") opens like a detective story or a mystery novel, with a puzzling and bizarre incident that reflects the changes that are starting to happen, slowly, stealthily, in his own brain. Karinthy himself would be both subject and investigator in the increasingly complex drama that he was subsequently drawn into.

Gifted and precocious (he had written his first novel at fifteen), Karinthy achieved fame in 1912, at the age of twenty-five, when no fewer than five of his books were published. Though he was trained in mathematics and actively interested in all aspects of science, he was especially known for his satirical writings, his political passions, and his surreal sense of humor. He had written philosophical works, plays, poems, novels, and, at the time of his first symptoms, had started writing a vast encyclopedia, which he hoped might be the twentieth-century equivalent of Diderot's monumental *Encyclopedia*. With all of this previous work, there had always been a plan, a structure, but now, forced to pay attention to what was happening in his own brain, Karinthy could only record, make notes, and reflect, without any clear notion of what lay ahead, of where this new journey would take him.

The hallucinatory train noises soon became a fixture in Karinthy's life. He started to hear them regularly, at seven o'clock each evening, whether he was in his favorite café or anywhere else. And within a few days, even stranger events started to occur:

> The mirror opposite me seemed to move. Not more than an inch or two, then it hung still. . . . But what was happening now? . . . I had no headache nor pain of any kind, I heard no trains, my heart was perfectly normal. . . . And yet everything, myself included, seemed to have lost its grip on reality. The tables remained in their usual places, two men were just walking across the café, and in front of me I saw the familiar water-jug and match-box. Yet in some eerie and alarming way they had all become accidental, as if they happened to be where they were purely by chance, and might just as well be anywhere else. . . . And now the whole box of tricks was starting to roll about, as if the floor underneath it had given way. I wanted to cling on to something. . . . There wasn't a fixed point anywhere. . . . Unless, perhaps, I could find one in my own head. If I could catch hold of a single image or memory or association that would help me to recognize myself. Or even a word might do.

This is a remarkable description of what it feels like to have the very foundations of perception, of consciousness, of self, undermined—to descend (perhaps for a few moments, but they may seem an eternity) into what Proust called "the abyss of unbeing," and to long desperately for some image, some memory, some word with which to pull oneself out.

At this point Karinthy started to realize that something might be seriously and strangely the matter; he wondered if he was having seizures or working up to a stroke. In the weeks that followed, he started to get further symptoms: attacks of retching and nausea, difficulties with balance and gait. He did his best to dismiss and discount these, but finally, concerned by a steadily increasing blurring of his vision, he consulted an ophthalmologist, and started on a frustrating medical odyssey:

> The doctor whom I called to consult shortly afterwards did not even examine me. Before I could describe half my symptoms he lifted his hand: "My dear fellow, you've neither aural catarrh nor have you had a stroke. . . . Nicotine poisoning, that's what's the matter with you."

WERE DOCTORS IN BUDAPEST in 1936 worse than doctors in, say, New York or London seventy years later? Not listening, not examining, being opinionated, jumping to conclusions—all are as ubiquitous, and dangerous, now as they were then and there (as Jerome Groopman describes so well in his book *How Doctors Think*). Wholly treatable disorders can go unrecognized, undiagnosed, until it is too late. Had Karinthy's first doctor examined him, he would have found a disorder of coordination indicating a cerebellar disturbance; looking into the eyes, he would have seen papilloedema—a swelling of the papillae, the optic disks—a sure sign of increased pressure in the brain.

Had he paid attention to what his patient tried to tell him, he would never have been so cavalier: no one has such auditory hallucinations or sudden underminings of consciousness without a significant cerebral cause.

But Karinthy was a part of the rich and fertile café culture of Budapest, and his social circle included not only writers and artists but scientists and doctors, too. This may have made it difficult for him to get a straightforward medical opinion, for his doctors were also his friends or colleagues. As the weeks passed, Karinthy, though making light of his symptoms, started to be haunted by two memories: that of a young friend who had died of a brain tumor, and that of a film he had once seen, showing the great pioneer neurosurgeon Harvey Cushing operating on the brain of a conscious patient.

At this point, suspecting that he, too, might have a brain tumor, Karinthy insisted that the ophthalmologist, a friend of his, examine his retinas closely. His vivid recounting of this scene is both shocking and richly ironic, and shows his sharp eye and comic gifts at their best. Taken aback a little at Karinthy's insistence, the doctor who had jollied him along a few months earlier now pulled out his ophthalmoscope and looked:

> As he bent close over me, I felt the ingenious little instrument brushing my nose and I could hear him draw his breath with a slight effort, as he strained to observe me closely. I waited for the usual reassurance. "Nothing wrong there! You just want new glasses—a trifle stronger this time. . . ." The reality was very different. I heard Dr. H. give a sudden whistle. . . .
>
> He laid down his instrument on the table and tilted his

head on one side. I saw him look at me with a kind of grave amazement, as if I had suddenly become a stranger to him.

Suddenly Karinthy ceased to be himself, a social acquaintance, an equal, a fellow human being with fears and feelings—and became a specimen. Dr. H. "was as pleasantly excited as an entomologist who has stumbled on some coveted specimen." He ran out of the room to summon his colleagues:

In an incredibly short space of time the room was full. Assistants, house physicians, students, came pouring round, greedily snatching the ophthalmoscope from one another.

The Professor himself came, turned to Dr. H., and said, "My congratulations! A really admirable diagnosis!"

As the medical men were congratulating one another, Karinthy tried to speak:

"Gentlemen . . . !" I began modestly.

Every one swung round. It was as if they had only just realized that I was of the party, and not only my papilla, which had become the centre of interest.

THIS SCENE IS ONE that could occur, and does occur, in hospitals all over the world—the sudden focus on an intriguing pathology, and the complete forgetting of the (perhaps terrified) human being who happens to have it. All doctors are guilty of this, which is why we continue to need books from the vantage point of patients. It is salutary to be reminded by a patient as witty and observant and articulate as Karinthy of how easily the human element is apt to be forgotten in the raptures of such an "entomological" excitement.

But one needs to remember, too, how difficult and delicate an art it was, seventy years ago, to diagnose and locate a cerebral tumor. In the 1930s, there were no MRI or CT scans, only elaborate and sometimes dangerous procedures, such as injecting air into the ventricles of the brain or injecting a dye into its blood vessels.

So it took months as Karinthy was referred from one specialist to another, and meanwhile his vision was growing worse. As he approached virtual blindness, he entered a strange world, where he could no longer be certain whether he was actually seeing or not:

> I had learnt to interpret every hint afforded by the shifting of light and to complete the general effect from memory. I was getting used to this strange semi-darkness in which I lived, and I almost began to like it. I could still see the outline of figures fairly well, and my imagination supplied the details, like a painter filling an empty frame. I tried to form a picture of any face I saw in front of me by observing the person's voice and movements. . . . The idea that I might already have gone blind struck me with sudden terror. What I fancied I saw was perhaps no more than the stuff that dreams are made on. I might only be using people's words and voices to reconstruct the lost world of reality. . . . I stood on the threshold of reality and imagination, and I began to doubt which was which. My bodily eye and my mind's eye were blending into one.

Just as Karinthy was on the verge of permanent blindness, a precise diagnosis of the tumor was made, finally, by the eminent Viennese neurologist Otto Pötzl, who recommended immediate surgery. Karinthy, accompanied by his wife, took a series of trains to Sweden to meet the great Herbert Olivecrona, a student of Harvey Cushing's and one of the best neurosurgeons in the world.

Karinthy's portrait of Olivecrona is full of insight and irony, and written in a new, spare style, quite different from the lush description that precedes it. The courtesy and reserve of the cool Scandinavian neurosurgeon is delicately brought out, in contrast to the Central European emotionalism of his illustrious patient. Karinthy is done with his ambivalence, his denials, his suspicions, and he has at last found a doctor whom he can trust and even love.

OLIVECRONA TELLS HIM that the operation will last many hours but that only local anesthetic will be used, because the brain itself has no sensory nerves, does not feel pain—and general anesthesia for such a lengthy operation is too risky. And, he adds, some parts of the brain, while not sensitive to pain, may, when stimulated, evoke vivid visual or auditory memories.

Karinthy described the initial drilling:

There was an infernal scream as the steel plunged into my skull. It sank more and more rapidly through the bone, and the pitch of its scream became louder and more piercing every second. . . . Suddenly, there was a violent jerk, and the noise stopped.

Karinthy heard a rush of fluid inside his head and wondered if it was blood or spinal fluid. He was then wheeled into the X-ray room, where air was injected into the ventricles of the brain to outline them and delineate the way in which they were being compressed by his tumor.

Back in the operating room, Karinthy was immobilized, face-down, on the operating table, and the surgery began in earnest. The greater part of his skull was exposed, and then much of it removed, piecemeal. Karinthy felt

> a straining sensation, a feeling of pressure, a cracking sound, and a terrific wrench. . . . Something broke with a dull noise. . . . This process was repeated many times . . . like splitting open a wooden packing-case, plank by plank.

Once the skull had been opened, all pain ceased—and this itself was paradoxically disturbing:

> No, my brain did not hurt. Perhaps it was more exasperating this way than if it had. I would have preferred it to hurt me. More terrifying than any actual pain was the fact that my position seemed impossible. It was impossible for a man to be lying here with his skull open and his brain exposed to the outer world—impossible for him to lie here and live . . . impossible, incredible, indecent, for him to remain alive—and not merely alive, but conscious and in his right mind.

At intervals, the cool, kindly voice of Olivecrona broke in, explaining, reassuring, and Karinthy's apprehension was replaced by calm and curiosity. Olivecrona, here, seems almost like Virgil, guiding his poet-patient through the circles and land-scapes of his brain.

Six or seven hours into the operation, Karinthy had a singular experience. It was not a dream, for he was fully conscious—though, perhaps, in an altered state of consciousness. He seemed to be looking down on his body from the ceiling of the operating theater, moving about, zooming in and out:

The hallucination consisted in my mind seeming to move freely about the room. There was only a single light, which fell evenly on to the table. Olivecrona . . . seemed to be lean-ing forward . . . the lamp on his forehead threw a light into the open cavity of my skull. He had already drained off the yellowish fluid. The lobes of the cerebellum seemed to have subsided and fallen apart of themselves, and I fancied I saw the inside of the opened tumour. He had cauterized the sev-ered veins with a red-hot electric needle. The angioma [the tumor made up of blood vessels] was already visible, lying within the cyst and a little to one side of it. The tumour itself looked like a great, red globe. In my vision it seemed as large as a small cauliflower. Its surface was embossed so that it formed a kind of pattern, like a cameo with a carved design. . . . It seemed almost a pity that Olivecrona was to destroy it.

Karinthy's visualization or hallucination continued in minute detail. He "saw" Olivecrona skillfully removing the tumor, suck-ing his lower lip with concentration, and then with satisfaction that the essential part of the surgery was done.

I do not know what to call this intense visualization, informed and conjured up by his detailed knowledge of what was actually happening. Karinthy himself uses the word "hallucination," and the aerial viewpoint, looking down on one's own body, is very characteristic of what is often called an "out-of-body experience." (Such OBEs are often associated with near-death experiences such as cardiac arrest or the perception of imminent catastrophe—and they have been associated with temporal lobe seizures and the stimulation of the temporal lobes during brain surgery.)

Whatever it was, Karinthy seemed to know that the operation had been successful, that the tumor had been removed without any damage to his brain. Perhaps Olivecrona had said this to him and Karinthy had transformed his words into a vision. After this intense and reassuring experience, Karinthy fell deeply asleep and did not wake up until he was back in his hospital bed.

The surgery, in Olivecrona's masterly hands, had gone well—the tumor, which turned out to be benign, was gone, and Karinthy made a complete recovery, even recovering his vision, which the doctors had thought would be permanently lost. He could read and write once again, and with an exuberant sense of relief and gratitude, he rapidly composed *A Journey Round My Skull,* and sent the first copy of the German edition to the surgeon who had saved his life. He followed this with another book, *The Heavenly Report,* somewhat different in style and approach, and then started on yet another, *Message in the Bottle.* He was apparently in full health and full creative swing when he died suddenly in August 1938. He was only fifty-one. It is said that he had a stroke while bending to tie his shoelace.

# Clinical Tales

# Cold Storage

In 1957, when I was a medical student under Richard Asher, I encountered his patient "Uncle Toby" and was fascinated by this strange meeting of fact and fable. Dr. Asher sometimes referred to it as a "Rip van Winkle case." The story often came to my mind, vividly, when my own postencephalitic patients were "awakened" in 1969, and it has unconsciously haunted me for years.

Dr. Asher had been on a house call to see a sick child. As he was discussing her treatment with the family, he noticed a silent, motionless figure in a corner.

"Who's that?" he asked.

"That's Uncle Toby—he's hardly moved in seven years."

Uncle Toby had become an undemanding fixture in the house. His slowing down was so gradual at first that the family didn't notice; but then, when it became more profound, it was—rather extraordinarily—just accepted by the family. He was fed and watered daily, turned, sometimes toileted. He was really no trouble; he was part of the furniture. Most people never noticed him, still, silent in the corner. He was not regarded as ill; he had just come to a stop.

Dr. Asher spoke to this waxlike figure. There was no answer, no response. He put out his hand to take the pulse and encountered a hand cold to the touch, almost as cold as that of a corpse. But there was a faint, slow pulse: Uncle Toby was alive, suspended, apparently, in some strange icy stupor.

Discussion with the family was odd and disquieting. They showed remarkably little concern for Uncle Toby, and yet, manifestly, they were caring and decent. Evidently, as sometimes happens with an insidious and insensible change, they had accommodated to it as it had happened. But when Dr. Asher spoke to them, and suggested that Uncle Toby be brought into the hospital, they agreed.

And so Uncle Toby was admitted to the hospital, to a specially equipped metabolic care unit, which is where I encountered him. His temperature could not be measured by an ordinary clinical thermometer, so a special one, reserved for hypothermics, was fetched; it registered sixty-eight degrees Fahrenheit. Uncle Toby's temperature was thirty degrees below normal. A suspicion was formed, immediately tested and confirmed: Uncle Toby had virtually no thyroid function, and his metabolic rate was reduced almost to zero. With scarcely any thyroid function, any metabolic stimulator or "fire," he had sunk into the depths of a hypothyroid (or myxedema) coma: alive but not alive; in abeyance, in cold storage.

It was clear what to do—it was a simple medical problem: we had only to give him a thyroid hormone, thyroxine, and he would come to. But this warming up, this refiring of metabolism, would have to be done very cautiously and slowly; his functions and his organs had accommodated to his hypometabolism. If his metabolism was stimulated too quickly, he might have cardiac or other complications. So slowly, very slowly, we started him on thyroxine, and very slowly he started to warm up . . .

A week passed. There was nothing to see, though Uncle Toby's temperature was now seventy-two degrees. It was only in the third week, with his body temperature now well over eighty degrees, that he began to move . . . and talk. His voice was exceedingly low, slow, and hoarse—like a phonograph record croaking round at a single revolution per minute. (Some of this croakiness was due to myxedema of the vocal cords.) His limbs, too, had been stiff and swollen with edema, but grew lither and more limber now with physiotherapy and use. After a month, though still cool, and slow in speech and motion, Uncle Toby had clearly "awakened," and he evinced animation, awareness, and concern.

"What's happening?" he asked. "Why am I in hospital? Am I ill?" We countered by asking him what *he* had been feeling. "Sort of cool, sort of lazy, slowed down, you know."

"But Mr. Oakins," we said—we called him "Uncle Toby" only among ourselves—"what happened in between feeling cool, feeling slow, and finding yourself here?"

"Nothing much," he answered. "Nothing I know of. I suppose I must have been really ill, passed out, and the family brought me here."

"And how long had you passed out for?" we asked, in a neutral tone.

"How long? A day or two—couldn't be any longer—my family would be sure to bring me in."

He scanned our faces curiously, intently.

"There's nothing more to this, nothing unusual?"

"Nothing," we reassured him, and made a quick exit.

. . .

MR. OAKINS, it seemed, unless we misunderstood him, had no sense that any time had elapsed, certainly not any great length of time. He had felt queer; now he was better—simple, nothing to it. Could this be what he actually believed?

We were given vivid confirmation of this later that same day, when the staff nurse came to us in some agitation. "He's quite lively now," she reported. "He has a real need to talk—he's talking about his mates, his work. About Attlee, the king's illness, the 'new' Health Service, and so on. He's no idea what's going on *now*. He seems to think it's 1950."

Uncle Toby, as a person, a conscious entity, had slowed down and stopped, as if he had gone into a coma. He had been "away," "absent," for an unconscionable time. Not in a sleep, not in a trance, but deeply submerged. And now that he had emerged, those years were a blank. It was not amnesia, not "disorientation"; his higher cerebral functions, his mind, had been "out" for seven years.

How would he react to the knowledge that he had lost seven years, and that much of what was exciting, important, dear to him had passed irretrievably away? That he himself was no longer contemporary but a piece of the past, an anachronism, a fossil strangely preserved?

Rightly or wrongly, we decided on a policy of evasion (and not only evasion but frank deceit). This was planned, of course, as a temporary measure, until he had the physical and mental strength to come to terms with things, to withstand a profound shock.

The medical staff made no efforts, therefore, to disabuse him of his belief that it was 1950. We watched ourselves closely, lest we give anything away; we forbade any careless talk; and we inundated him with newspapers and periodicals from 1950. He read these avidly, though he expressed surprise, on occasion, at *our* ignorance of the "news," as well as the disgraceful, yellowed, dilapidated condition of the papers.

And now—six weeks had passed—his temperature was almost normal. He looked fit and well, and considerably younger than his years.

At this point came the final irony. He started to cough, to spit blood; he had a massive hemoptysis. Chest X-rays showed a mass in his chest, and bronchoscopy revealed a highly malignant, rapidly proliferating oat-cell carcinoma.

We managed to find chest films, routine X-rays, from 1950, and there we saw, small and overlooked at the time, the cancer he now had. Such highly malignant, fulminating carcinomas are apt to grow rapidly and be fatal in months—yet he had had this for seven years. It seemed evident that the cancer, like the rest of him, had been arrested, in cold storage. Now that he was warmed up, the cancer raged furiously, and Mr. Oakins expired, in a fit of coughing, a matter of days later.

His family let him sink into coldness, which saved his life; we warmed him up, and, in consequence, he died.

# Neurological Dreams

However dreams are to be interpreted—the Egyptians saw them as prophecies and portents; Freud as hallucinatory wish fulfillments; Francis Crick and Graeme Mitchison as "reverse learning" designed to remove overloads of "neural garbage" from the brain—it is clear that they may also contain, directly or distortedly, reflections of current states of body and mind.

Thus it is scarcely surprising that neurological disorders—in the brain itself or in its sensory or autonomic input—can alter dreaming in striking and specific ways. Every practicing neurologist must be aware of this, and yet we rarely question patients about their dreams. There is virtually nothing on this subject in the medical literature, but I think such questioning can be an important part of the neurological examination, can assist in diagnosis, and can show how sensitive a barometer dreaming may be of neurological health and disease.

I first encountered this many years ago while working in a migraine clinic. It became clear that there was not only a general correlation between the incidence of very intense dreams or nightmares and visual migraine auras but also, not infrequently, an entering of migraine aura phenomena into the dreams. Patients might dream of phosphenes or zigzags, of expanding scotomas or of colors or contours that wax and then fade. Their dreams might contain visual field defects or hemianopia or, more rarely, the phenomena of "mosaic" or "cinematic" vision.

The neurological phenomena in such cases may appear direct and raw, intruding into the otherwise normal unfolding of a dream. But they may also combine with the dream, fuse with and be modified by its images and symbols. Thus the phosphenes of migraine are often dreamed of as fireworks displays, and one patient of mine sometimes embedded his nocturnal migraine auras in dreams of a nuclear explosion. He would first see a dazzling fireball with a typically migrainous, iridescent zigzag margin, coruscating as it grew, until it was replaced by a blind area (or scotoma) with the dream round its edge. At this point he would usually wake with a fading scotoma, intense nausea, and an incipient headache.

If there are lesions in the occipital, or visual, cortex, patients may observe specific visual deficits in their dreams. Mr. I., the colorblind painter I described in *An Anthropologist on Mars,* had a central achromatopsia, and he remarked that he no longer dreamed in color. People with certain types of prestriate lesions may, while dreaming, be unable to recognize faces, a condition called prosopagnosia. And one patient of mine, with an angioma in his occipital lobe, knew that if his dreams were suddenly suffused with a red color, if they "turned red," he was in for a seizure. If the damage to the occipital cortex is diffuse enough, visual imagery may vanish completely from dreams. I have encountered this, on occasion, as a presenting symptom of Alzheimer's disease.

. . .

ANOTHER PATIENT, a man who had focal sensory and motor seizures, dreamed that he was in court, being prosecuted by Freud, who kept banging on his head with a gavel as the charges were being read. But the blows, strangely, were felt in his left arm, and he awoke to find it numb and convulsing, in a typical focal seizure.

The most common neurological or "physical" dreams are of pain, discomfort, hunger, or thirst, at once manifest and yet camouflaged in the scenery of the dream. Thus one patient, newly casted after a leg operation, dreamed that a heavy man had stepped, with agonizing effect, on his left foot. Politely at first, then with increasing urgency, he asked the man to move, and when his appeals were unheeded, he tried to shift the man bodily. His efforts were completely useless, and now, in his dream, he realized why: the man was made of compacted neutrons—neutronium—and weighed six trillion tons, as much as the earth. He made one last, frenzied attempt to move the immovable, then woke up with an intense viselike pain in his foot, which had become ischemic from the pressure of the new cast.

Patients may sometimes dream of the onset of a disease before it physically manifests. A woman whom I described in *Awakenings* was stricken with acute encephalitis lethargica in 1926 and had a night of grotesque and terrifying dreams about one central theme: she dreamed she was imprisoned in an inaccessible castle, but the castle had the form and shape of herself. She dreamed of enchantments, bewitchments, entrancements; she dreamed she had become a living sentient statue of stone;

she dreamed the world had come to a stop; she dreamed she had fallen into a sleep so deep that nothing could wake her; she dreamed of a death that was different from death. Her family had difficulty waking her the next morning, and when she awoke there was intense consternation: overnight, she had become parkinsonian and catatonic.

Christina, a woman I described in *The Man Who Mistook His Wife for a Hat,* was admitted to the hospital before surgery to remove her gallbladder. She was placed on antibiotics for microbial prophylaxis; since she was an otherwise healthy young woman, no complications were expected. The night before surgery, though, she had a disturbing dream of peculiar intensity. She was swaying wildly, in the dream, very unsteady on her feet, could hardly feel the ground beneath her, could hardly feel anything in her hands, found them flailing to and fro, kept dropping whatever she picked up.

She was distressed by this dream ("I never had one like it," she said. "I can't get it out of my mind")—so distressed that we requested an opinion from the psychiatrist. "Preoperative anxiety," he said. "Quite natural, we see it all the time." But within a few hours the dream had become a reality, as the patient became incapacitated by an acute sensory neuropathy—she had lost the sense of proprioception and could no longer tell where her limbs were without looking. One must assume in such a case that the disease was already affecting her neural function and that the unconscious mind, the dreaming mind, was more sensitive to this than the waking mind. Such premonitory or precursory dreams may sometimes be happy in content and in outcome, too. Patients with multiple sclerosis may dream of remissions a few hours before they occur, and patients recovering from strokes or neurological injuries may have striking dreams of improvement before such improvement is objectively manifest. Here again, the dreaming mind may be a more sensitive indicator of neural function than examination with a reflex hammer and a pin.

Some dreams seem to be more than precursory. One striking personal example (which I described at length in *A Leg to Stand On*) stays in my mind. While recovering from a leg injury, I had been told it was time to advance from using two crutches to just one. I tried this twice, and both times fell flat on my face. I could not consciously think how to do it. Then I fell asleep and had a dream in which I reached out my right hand, grabbed the crutch that hung over my head, tucked it under my right arm, and set off with perfect confidence and ease down the corridor. Waking from the dream, I reached out my right hand, grabbed the crutch that hung over the bed, and set off with perfect confidence and ease down the corridor.

This, it seemed to me, was not merely premonitory but a dream that actually did something, a dream that solved the very motor-neural problem the brain was confronted with, achieving this in the form of a psychic enactment or rehearsal or trial: a dream, in short, that was an act of learning.

Disturbances in body image from limb or spinal injuries almost always enter dreams, at least when they are acute, before any "accommodation" has been made. With my own deafferenting leg injury, I had reiterative dreams of a dead or absent limb. Within a few weeks, however, such dreams tend to cease, as there occurs a revision or "healing" of body image in the cortex. (Such changes in cortical mapping have been demonstrated in Michael Merzenich's experiments with monkeys.) Phantom limbs, by contrast, perhaps because of continuing neural excitation in the stump, intrude themselves into dreams (as into waking consciousness) very persistently, though gradually telescoping and growing fainter with the passage of years.

The phenomena of parkinsonism may also enter dreams. Ed W., a man of acute introspective ability, felt that the first expression of parkinsonism in him was a change in the style of his dreams. He would dream that he could move only in slow motion, or that he was "frozen," or that he was rushing and could not stop. He would dream that space and time themselves had changed, kept "switching scales," and had become chaotic and problematic. Gradually, over the ensuing months, these looking-glass dreams came true, and his bradykinesia and festination became obvious to others. But the symptoms had first presented themselves in his dreams.[1]

Alterations in dreaming are often the first sign of response to L-dopa in patients with ordinary Parkinson's disease, as well as in those with postencephalitic parkinsonism. Dreaming typically becomes more vivid and more emotionally charged (many patients remark that they are dreaming, suddenly, in brilliant color). Sometimes the realness of these dreams is so extraordinary that they cannot be forgotten or thrown off after waking.

Excessive dreaming of this sort, excessive both in sensory vividness and in activation of unconscious psychic content—dreaming akin, in some ways, to hallucinosis—is common in fever or delirium, as a reaction to many drugs (opi-

---

1. Another man I know, one who has Tourette's syndrome, felt that he frequently had "Touretty" dreams—dreams of a particularly wild and exuberant kind, full of unexpectednesses, accelerations, and sudden tangents. This changed when he was put on a tranquilizer, haloperidol; he then reported that his dreams had been reduced to "straight wish fulfillment, with none of the elaboration, the extravagances of Tourette's."

ates, cocaine, amphetamines, and so on), and in states of drug withdrawal or REM rebound. A similar unbridled oneirism may occur at the start of some psychoses, where an initial mad or manic dream, like the rumbling of a volcano, may be the first intimation of the eruption to come.

Dreaming, for Freud, was the "royal road" to the unconscious. Dreaming, for the physician, may not be a royal road, but it is a byway to unexpected diagnoses and discoveries, and to unexpected insights about how one's patients are doing. It is a byway full of fascination, and should not be neglected.

# Nothingness

Nature abhors a vacuum—and so do we. The idea of a void—of emptiness, nothingness, spacelessness, placelessness, all such "lessness"—is at once abhorrent and inconceivable, and yet it haunts us in the strangest, most paradoxical way. As Beckett writes, "Nothing is more real than nothing."

For Descartes there was no such thing as empty space. For Einstein there was no space without field. For Kant our ideas of space and extension were the forms our "reason" gives to experience, through the operation of a universal "synthetic a priori." The nervous system, intact and active, was envisaged by Kant as a sort of transformer, forming ideality from reality, reality from ideality. Such a notion has the virtue—very rare in metaphysical formulations—that it can instantly be tested in practice; specifically, in neurological and neurophysiological practice.

If one is given a spinal anesthetic that brings to a halt neural traffic in the lower half of the body, for instance, one cannot feel merely that this is paralyzed and senseless; one feels that it is wholly, impossibly, "nonexistent," that one has been cut in half, and that the lower half is absolutely missing—not in the familiar sense of being somewhere, elsewhere, but in the uncanny sense of *not-being*, or being nowhere. Patients under spinal anes-thesia may say that part of them is "missing" or "gone," that it seems like dead flesh, or sand, or paste; devoid of life, of "will." One such patient, trying to formulate the unformulable, finally said that his lost limbs were "nowhere to be found" and that they were "like nothing on earth." Hearing such phrases, one is reminded of the words of Hobbes: "That which is not Body, is no part of the Universe: And because the Universe is All, that which is not part of it, is *Nothing*; and consequently *no where.*"

Spinal anesthesia provides a striking and dramatic example of a *transient* "annihilation," but there are many simpler exam-ples of annihilation in everyday life. All of us have sometimes slept on an arm, crushing its nerves and briefly extinguishing neural traffic; the experience, though very brief, is an uncanny one because our arm seems to be no longer "ours," but an inert, senseless nothing that is not part of ourselves. Wittgenstein grounds "certainty" in the certainty of the body: "If you can say, *here is one hand,* we'll grant you all the rest." But when you wake up after nerve-crushing your arm, you cannot say, "This is my hand" or even "This is a hand," except in a purely formal sense. What has always been taken for granted, or axiomatic, is revealed as radically precarious and contingent; having a body, having *anything,* depends on one's nerves.

There are countless other situations—physiological and path-ological, common or uncommon—in which there are brief, or prolonged, or permanent annihilations. Strokes, tumors, injuries, especially to the right half of the brain, tend to cause a partial or total annihilation of the left side—a condition variously known as "imperception," "inattention," "neglect," "agnosia," "anosog-nosia," "extinction," or "alienation." All of these are experiences of nothingness (or, more precisely, privations of the experience of somethingness).

Blockage to the spinal cord or the great limb plexuses can produce a similar situation, even though the brain is intact, for it is deprived of the information from which it might form an image (or a Kantian "intuition"). Indeed it can be shown by measuring electrical potentials in the brain during spinal or regional blocks that there is a dying away of activity in the corresponding part of the cerebral representation of the "body image"—the empirical reality required for Kantian ideality. Similar annihilations may be brought out peripherally, either through nerve or muscle damage in a limb or by simply enclosing the limb in a cast, which by its mixture of immobilization and encasement may temporarily bring neural traffic and impulses to a halt.

Nothingness, annihilation, is a reality in this ultimately paradoxical sense.

# Seeing God in the
# Third Millennium

There are many carefully documented accounts in the medical literature of intense, life-altering religious experiences during epileptic seizures. Hallucinations of overwhelming intensity, sometimes accompanied by a sense of bliss and a strong feeling of the numinous, can occur, especially with the so-called ecstatic seizures that may be brought on by temporal lobe epilepsy.[1] Though such seizures may be brief, they can lead to a fundamental reorientation, a metanoia, in one's life. Fyodor Dostoevsky was prone to such seizures and described many of them, including this one:

> The air was filled with a big noise and I tried to move. I felt the heaven was going down upon the earth and that it engulfed me. I have really touched God. He came into me myself, yes God exists, I cried, and I don't remember anything else. You all, healthy people . . . can't imagine the hap-

---

1. I have described ecstatic seizures, as well as near-death experiences, at greater length in *Hallucinations*.

piness which we epileptics feel during the second before our fit. . . . I don't know if this felicity lasts for seconds, hours or months, but believe me, for all the joys that life may bring, I would not exchange this one.

A century later, Kenneth Dewhurst and A. W. Beard published a detailed report in the *British Journal of Psychiatry* of a bus conductor who had a sudden feeling of elation while collecting fares. They wrote:

He was suddenly overcome with a feeling of bliss. He felt he was literally in Heaven. He collected the fares correctly, telling his passengers at the same time how pleased he was to be in Heaven. . . . He remained in this state of exaltation, hearing divine and angelic voices, for two days. Afterwards he was able to recall these experiences and he continued to believe in their validity. [Three years later,] following three seizures on three successive days, he became elated again. He stated that his mind had "cleared." . . . During this episode he lost his faith.

He now no longer believed in heaven and hell, in an afterlife, or in the divinity of Christ. This second conversion—to atheism—carried the same excitement and revelatory quality as the original religious conversion.

More recently, Orrin Devinsky and his colleagues have been able to make video EEG recordings in patients who are having such seizures, and have observed an exact synchronization of the epiphany with a spike in epileptic activity in the temporal lobes (more commonly the right temporal lobe).

Ecstatic seizures are rare—they only occur in something like 1 or 2 percent of patients with temporal lobe epilepsy. But the last half century has seen an enormous increase in the prevalence of other states sometimes permeated by religious joy and awe, "heavenly" visions and voices, and, not infrequently, religious conversion or metanoia. Among these are out-of-body experiences (OBEs)—which are more common now that more patients can be brought back to life from serious cardiac arrests and the like—and much more elaborate and numinous experiences called near-death experiences (NDEs).

Both OBEs and NDEs, which occur in waking but often profoundly altered states of consciousness, cause hallucinations so vivid and compelling that those who experience them may deny the term "hallucination" and insist on their reality. And the fact that there are marked similarities in individual descriptions is taken by some to indicate their objective "reality."

But the fundamental reason that hallucinations—whatever their cause or modality—seem so real is that they deploy the very same systems in the brain that actual perceptions do. When one hallucinates voices, the auditory pathways are activated; when one hallucinates a face, the fusiform face area, normally used to perceive and identify faces in the environment, is stimulated.

In OBEs, subjects feel that they have left their bodies—they seem to be floating in midair or in a corner of the room, looking down on their vacated bodies from a distance. The experience may be felt as blissful, terrifying, or neutral. But its extraordinary nature, the apparent separation of "spirit" from body, imprints it indelibly on the mind and may be taken by some people as evidence of an immaterial soul—proof that consciousness, personality, and identity can exist independently of the body and even survive bodily death.

Neurologically, OBEs are a form of bodily illusion arising from a temporary dissociation of visual and proprioceptive representations—normally these are coordinated, so that one views the world, including one's body, from the perspective of one's own eyes, one's head. OBEs, as Henrik Ehrsson and his fellow researchers in Stockholm have elegantly shown, can be produced experimentally, by using simple equipment—video goggles, mannequins, rubber arms, and so on—to confuse one's visual input and one's proprioceptive input and create an uncanny sense of disembodiedness.

A number of medical conditions can lead to OBEs—cardiac arrest or arrhythmias or a sudden lowering of blood pressure or blood sugar, often combined with anxiety or illness. I know of some patients who have experienced OBEs during difficult childbirths, and others who have had them in association with narcolepsy or sleep paralysis. Fighter pilots subjected to high g-forces in flight (or sometimes in training centrifuges) have reported OBEs as well as much more elaborate states of consciousness that resemble a near-death experience.

The near-death experience usually goes through a sequence of characteristic stages. One seems to be moving effortlessly and blissfully along a dark corridor or tunnel towards a wonderful "living" light—often interpreted as heaven or the boundary between life and death. There may be a vision of friends and relatives welcoming one to the other side, and there may be a rapid yet extremely detailed series of memories of one's life—a lightning autobiography. The return to one's body may be abrupt, as when, for example, the beat is restored to an arrested heart. Or it may be more gradual, as when one emerges from a coma.

Not infrequently, an OBE turns into an NDE—as happened with Tony Cicoria, a surgeon who told me how he had been struck by lightning. He gave me a vivid account of what then followed, as I wrote in *Musicophilia*:

> "I was flying forwards. Bewildered. I looked around. I saw my own body on the ground. I said to myself, 'Oh shit, I'm dead.' I saw people converging on the body. I saw a woman—she had been standing waiting to use the phone right behind me—position herself over my body, give it CPR. . . . I floated up the stairs—my consciousness came with me. I saw my kids, had the realization that they would be okay. Then I was surrounded by a bluish-white light . . . an enormous feeling of well-being and peace. The highest and lowest points of my life raced by me . . . pure thought, pure ecstasy. I had the perception of accelerating, being drawn up . . . there was speed and direction. Then, as I was saying to myself, 'This is the most glorious feeling I have ever had'—SLAM! I was back."

Dr. Cicoria had some memory problems for a month or so after this, but he was able to resume his practice as an orthopedic surgeon. Yet he was, as he put it, "a changed man." Previously he had no particular interest in music, but now he was seized by an overwhelming desire to listen to classical music, especially Chopin. He bought a piano and started to play obsessively and to compose. He was convinced that the entire episode—being struck by lightning, having a transcendent vision, then being resuscitated and given this gift so that he could bring music to the world—was part of a divine plan.

Cicoria has a PhD in neuroscience, and he felt that his sudden accession of spirituality and musicality must have been connected to changes in his brain—changes we might be able to clarify, perhaps, with neuroimaging. He saw no contradiction between religion and neurology—if God worked on a man or in a man, Cicoria felt, He would do so via the nervous system, via parts of the brain specialized, or potentially specializable, for spiritual feeling and belief. Cicoria's reasonable and (one might say) scientific attitude to his own spiritual conversion is in marked contrast to that of another surgeon, Dr. Eben Alexander, who describes, in his book *Proof of Heaven: A Neurosurgeon's Journey into the Afterlife,* a detailed and complex NDE that occurred while he spent seven days in a coma caused by meningitis. During his NDE, he writes, he passed through the bright light—the boundary between life and death—to find himself in an idyllic and beautiful meadow (which he realized was heaven), where he met a beautiful but unknown woman who conveyed various messages to him telepathically. Advancing farther into the afterlife, he felt the ever-more-embracing presence of God. Following this experience, Alexander became something of an evangelist, wanting to spread the good news that heaven really exists.

Alexander makes much of his experience as a neurosurgeon and an expert on the workings of the brain. He provides an appendix to his book detailing "Neuroscientific Hypotheses I Considered to Explain My Experience"—but he dismisses all of these as inapplicable in his own case because, he insists, his cerebral cortex was completely shut down during the coma, precluding the possibility of any conscious experience.

Yet his NDE was rich in visual and auditory detail, as many such hallucinations are. He is puzzled by this, since such sensory details are normally produced by the cortex. Nonetheless, his consciousness had journeyed into the blissful, ineffable realm of the afterlife—a journey that, he felt, lasted for most of the time he lay in a coma. Thus, he proposes, his essential self, his "soul," did not need a cerebral cortex or, indeed, any material basis whatever.

It is not so easy, however, to dismiss neurological processes. Dr. Alexander presents himself as emerging from his coma suddenly: "My eyes opened . . . my brain . . . had just kicked back to life." But one almost always emerges gradually from a coma; there are intermediate stages of consciousness. It is in these transitional stages, where consciousness of a sort has returned, but not yet fully lucid consciousness, that NDEs tend to occur.

Alexander insists that his journey, which he believes lasted for days, could not have occurred except while he was deep in the coma. But we know from the experience of Tony Cicoria and many others that a hallucinatory journey to the bright light and beyond, a full-blown NDE, can occur in twenty or thirty seconds, even though it seems to last much longer. Subjectively, during such a crisis, the very concept of time may seem variable or meaningless. The one most plausible hypothesis in Dr. Alexander's case, then, is that his NDE occurred not during his coma, but as he was surfacing from the coma and his cortex was returning to full function. It is curious that he does not allow this obvious and natural explanation but instead insists on a supernatural one.

To deny the possibility of any natural explanation for an NDE, as Dr. Alexander does, is more than unscientific—it is anti-scientific. It precludes the scientific investigation of such states.

Kevin Nelson, a neurologist at the University of Kentucky, has studied the neural basis of NDEs and other forms of "deep" hallucinating for many decades. In 2011, he published a wise and careful book about his research, *The Spiritual Doorway in the Brain: A Neurologist's Search for the God Experience.*

Nelson feels that the "dark tunnel" described in most NDEs represents constriction of the visual fields due to compromised blood pressure in the eyes, and the "bright light" represents a flow of visual excitation from the brainstem, through visual relay stations, to the visual cortex (the so-called pons-geniculate-occipital pathway).

Simpler perceptual hallucinations—of patterns, animals, people, landscapes, music, and so on—as one may get in a variety of conditions (blindness, deafness, epilepsy, migraine, or sensory deprivation, for example) do not usually involve profound changes in consciousness and, while very startling, are nearly always recognized as hallucinations. It is different with the very complex hallucinations of ecstatic seizures or NDEs—which are often taken to be veridical, truth-telling, and often life-transforming revelations of a spiritual universe and, perhaps, of a spiritual destiny or mission.

The tendency to spiritual feeling and religious belief lies deep in human nature and seems to have its own neurological basis, though it may be very strong in some people and less developed in others. For those who are religiously inclined, an NDE may seem to offer "proof of Heaven," as Eben Alexander puts it.

Some religious people come to experience their proof of heaven by another route—the route of prayer, as the anthropologist T. M. Luhrmann has explored in her book *When God Talks Back*. The very essence of divinity, of God, is immaterial. God cannot be seen, felt, or heard in the ordinary way. Luhrmann wondered how, in the face of this lack of evidence, God becomes a real, intimate presence in the lives of so many evangelicals and other people of faith.

She joined an evangelical community as a participant-observer, immersing herself in particular in their disciplines of prayer and visualization—imagining in ever-richer, more concrete detail the figures and events depicted in the Bible. Congregants, she writes,

> practice seeing, hearing, smelling, and touching in their mind's eye. They give these imagined experiences the sensory vividness associated with the memories of real events. What they are able to imagine becomes more real to them.

Sooner or later, with this intensive practice, for some of the congregants, the mind may leap from imagination to hallucination, and the congregant hears God, sees God, feels God walking beside them. These yearned-for voices and visions have the reality of perception, and this is because they activate the perceptual systems of the brain, as all hallucinations do. These visions, voices, and feelings of "presence" are accompanied by intense emotion—emotions of joy, peace, awe, revelation. Some evangelicals may have many such experiences, others only a single one—but even a single experience of God, imbued with the overwhelming force of actual perception, can be enough to sustain a lifetime of faith. (For those who are not religiously inclined, such experiences may occur with meditation or intense concentration on an artistic or intellectual or emotional plane, whether this is falling in love or listening to Bach, observing the intricacies of a fern or cracking a scientific problem.)

In the last decade or two, there has been increasingly active research in the field of "spiritual neurosciences." There are special difficulties in this research, for religious experiences cannot be summoned at will; they come, if at all, in their own time and way—the religious would say in God's time and way. Nonetheless, researchers have been able to demonstrate physiological changes not only in pathological states like seizures, OBEs, and NDEs, but also in positive states like prayer and meditation. Typically these changes are quite widespread, involving not only primary sensory areas in the brain but limbic (emotional) systems, hippocampal (memory) systems, and the prefrontal cortex, where intentionality and judgment reside.

Hallucinations, whether revelatory or banal, are not of supernatural origin; they are part of the normal range of human consciousness and experience. This is not to say that they cannot play a part in the spiritual life or have great meaning for an individual. Yet while it is understandable that one might attribute value, ground beliefs, or construct narratives from them, hallucinations cannot provide evidence for the existence of any metaphysical beings or places. They provide evidence only of the brain's power to create them.

# Hiccups and Other Curious Behaviors

In *On the Move*, I recounted the story of a man I met in 1960, when I worked as a research assistant in San Francisco for Grant Levin and Bertram Feinstein, two neurosurgeons whose specialty was operating on patients with parkinsonism.

One of their patients, Mr. B., was a coffee merchant who had survived an attack of encephalitis lethargica during the great epidemic of the 1920s but was now very disabled by postencephalitic parkinsonism. Mr. B. was a little frail and had emphysema, but otherwise seemed an excellent candidate for a cryosurgery that had been developed to reduce parkinsonian tremor and stiffness.

Immediately after the procedure, though, he developed hiccups, a symptom which at first we took to be trivial and transient. But his hiccups did not go away; they grew stronger and stronger, spreading to muscles in his back and abdomen, jolting his entire trunk. They were so violent as to interfere with eating, and they made sleep almost impossible. We tried the usual remedies—breathing into a paper bag, and so on—but none of these worked.

After six days and nights of continued hiccupping, Mr. B. was exhausted and frightened—the more so since he had heard that hiccups, or their debilitating effects, could be fatal.

Hiccupping involves a sudden jerk of the diaphragm, and sometimes, as a last resort for intractable hiccups, surgeons may block the phrenic nerves that supply the diaphragm. But this means that diaphragmatic breathing is no longer possible—one can only breathe shallowly, using the intercostal muscles in the chest. This was not an option with Mr. B., for he had emphysema and could not have survived without the use of his diaphragm.

Hesitantly, I suggested hypnosis, and Levin and Feinstein, though skeptical, agreed that we had nothing to lose. We found a hypnotist and were astounded when he managed to induce a hypnotic state in Mr. B.—this in itself seemed little short of miraculous, given his constant hiccupping. The hypnotist planted a posthypnotic suggestion: "When I snap my fingers, you will wake up and no longer have hiccups." He let the exhausted man sleep for ten more minutes and then snapped his fingers. Mr. B. came to, looking slightly confused—but free of hiccups. There were no relapses, and Mr. B., much helped by the cryosurgery, lived for several more years.

MR. B. WAS AMONG the hundreds of thousands who survived the worldwide epidemic of "sleepy sickness"—the encephalitis lethargica that raged between 1917 and 1927—only to suffer, sometimes years later, from postencephalitic syndromes of various sorts. Encephalitis lethargica could produce a wide array of lesions affecting the hypothalamus, the basal ganglia, the midbrain, and the brainstem, while sparing the cerebral cortex for the most part. Thus it particularly affected control mechanisms in the subcortex—systems involved with the regulation of sleep, sexuality, appetite; of posture, balance, and movement; and, at the brainstem level, autonomic functions like the regulation of breathing. These control systems are of great phylogenetic antiquity—occurring in most vertebrates.[1]

Many postencephalitics went on to develop an extreme form of parkinsonism, and they were also apt to develop various odd respiratory behaviors. These were especially severe in the immediate aftermath of the epidemic, although they tended to diminish with the passage of years. There were even "epidemics" of postencephalitic hiccup in several places.

There could also be spontaneous sneezing, coughing, or yawning among the victims of the sleepy sickness, as well as paroxysmal laughter or crying. These are normal, if curious, behaviors, as Robert Provine emphasizes in his book *Curious Behavior: Yawning, Laughing, Hiccupping, and Beyond*. But they are rendered abnormal by their severity, their incessancy, and

---

1. Hiccups can appear in fetuses as early as eight weeks after gestation but diminish in the later stages of pregnancy. Though hiccups have no obvious function after birth, they may be a vestigial behavior, perhaps a vestige of the gill movements of our fishy ancestors. A similar thought can arise when one sees, in patients with certain brainstem lesions, synchronous movements affecting muscles in the neck, the palate, and the middle ear. These muscles seem to have little to do with each other until one realizes that they are all vestiges of the branchial or gill muscles of fishes—neurologists speak, therefore, of branchial myoclonus. (Many similar examples, both anatomical and functional, are discussed by Neil Shubin in *Your Inner Fish*.)

their occurrence in the absence of any demonstrable cause—such patients did not have irritation of the esophagus, the diaphragm, the throat or nostrils; they had nothing to laugh or cry about. Yet they were overcome by hiccups, coughs, sneezes, yawns, laughter, or crying, presumably due to lesions in the brain stimulating or releasing such behaviors so that they occurred in an autonomous and inappropriate fashion.[2]

By 1935, most of these postencephalitic patients were submerged in an all-embracing catatonia or deep parkinsonism, and their odd respiratory behaviors had all but disappeared.

THIRTY YEARS LATER, I was working at Beth Abraham Hospital in the Bronx with eighty-odd postencephalitic patients; and while most had parkinsonism and sleep disorders, none of them had the overt respiratory disorders described in the earlier literature. But this changed when I gave them L-dopa in 1969 and many subsequently developed respiratory and phonatory tics, including sudden deep breaths, yawns, coughs, sighs, grunts, and sniffs.

I asked each of these patients whether they had ever experienced such respiratory symptoms in the past. Most could not give me a clear answer, but Frances D., an intelligent and articulate woman, said that she had indeed had respiratory crises from 1919 (when she came down with encephalitis lethargica) to 1924, but not thereafter. It seemed probable, in her case, that L-dopa had activated or released a preexisting sensitivity or pro-

2. This may be analogous to the occurrence of the "forced" laughter or crying sometimes seen in multiple sclerosis, ALS, Alzheimer's disease, after some strokes, or in some patients with epilepsy who suffer so-called dacrystic (crying) or gelastic (laughing) seizures.

clivity to respiratory disorders, and I had to wonder whether this could have been the case with the other patients who developed respiratory symptoms.

I was reminded of Mr. B., the postencephalitic coffee merchant with hiccups. Could he, too, have had damaged and hypersensitive respiratory controls, which, in his case, were released by a surgical lesion to the basal ganglia?

There tended to be, with the continued use of L-dopa, an elaboration of these respiratory or phonatory behaviors—not only grunting and coughing but hooting and snorting, hissing and whistling, barking, bleating, lowing and mooing, humming and buzzing. Rolando O., as I wrote in *Awakenings,* would make a sort of "murmuring-purring sound emitted with each expiration, rather pleasing to the ear, like the sound of a distant sawmill, or bees swarming, or a contented lion after a satisfactory meal." (Smith Ely Jelliffe, writing in the 1920s at the height of the epidemic, spoke of "menagerie noises" in such postencephalitic patients. With an entire ward of such patients at Beth Abraham now activated by L-dopa, startled visitors to the hospital sometimes wondered if there was indeed a menagerie up on the fifth floor, where my patients resided.)

Further elaboration occurred in several patients—for Frank G., a humming noise became a verbigeration of the phrase "keep cool, keep cool," which he uttered hundreds of times a day. Other patients developed chanting tics—tics given a rhythmic, melodic form, with a word or phrase embedded in them.[3]

Once, doing late-night rounds among my postencephalitic patients, I heard a singular sound, a sort of chorus, from

3. In *Musicophilia* I described a similar evolution of an expiratory/phonatory tic to full-blown incantations, in a man with tardive dyskinesia ("Accidental Davening").

a four-bedded room. When I looked in, I found that all four patients were asleep but singing in their sleep—a rather dreary, repetitive singsong melody, but one in which the four voices were synchronized and attuned with each other. Sleepwalking, sleep talking, and sleep singing were not uncommon in these sleepy-sickness patients, but it was the *coordination* of the four sleeping singers that amazed me. I wondered whether it had started with Rosalie B., a very musical woman, and spread by a sort of contagion to the other sleepers.

A VAST NUMBER of other involuntary behaviors were activated or released with L-dopa, virtually every subcortical function taking on a life of its own, occurring autonomously and spontaneously but amplified by involuntary imitations and mimicries as the patients saw and heard each other.

Frances D. showed a disintegration of the normal automatic controls of breathing within ten days of starting L-dopa. Her breathing became rapid, shallow, and irregular, broken up by sudden violent inspirations. Within a few days these differentiated into clear respiratory crises that would start without any warning, with a sudden inspiratory gasp followed by forced breath-holding for ten or fifteen seconds, then a violent expiration. These attacks become more and more intense, lasting almost a minute, during which Frances would struggle to expel air through a closed glottis, in so doing becoming purple and congested from the futile effort; finally the breath would be expelled with tremendous force, making a noise like the boom of a gun.

I observed similar propensities in Frances's roommate Martha, who had rapid breathing and difficulty in catching her breath, moving towards full-blown respiratory crises. These women's symptoms were so similar that I had to wonder whether one of them was "imitating" the other, a thought that was reinforced when Miriam, a third patient in their four-bedded room, also started to get progressively more severe respiratory disorders:

> The first such effect [I noted] was hiccup, which would come in hour-long attacks at 6:30 every morning. . . . A "nervous" cough and throat-clearing started, associated with a recurrent tic-like feeling of something blocking or scratching her throat . . . [then] a tendency to gasping and breath-holding, which in turn "replaced" the throat-clearing and coughing . . . finally full-blown respiratory crises that closely resembled those of Miss D.

Another patient, Lillian W., had at least a hundred clearly different forms of crises: hiccups, panting attacks, oculogyric attacks, sniffing, sweating, chattering of the teeth, attacks in which her left shoulder would grow warm, and paroxysmal ticcing. She had ritualized iterative attacks, in which she would tap one foot in three different positions or dab her forehead in four set places; counting attacks; verbigerative attacks, in which certain phrases were said a certain number of times; fear attacks, giggling attacks, and so on. Any mention of a particular attack to Lillian would invariably bring it on. She was deeply suggestible, especially during her oculogyric crises.

It was common for all of these curious behaviors not only to persist but to build up in intensity and spread, as if the brain was becoming sensitized and conditioned, learning or becoming overwhelmed by these perverse behaviors. These behaviors have a life of their own and once started may have to run their course; they can be difficult to stop by an act of will. They connect us to the origins of vertebrate behavior, and the ancient core of the vertebrate brain—the brainstem.

# Travels with Lowell

In 1986, I met a young photojournalist, Lowell Handler, who told me he had Tourette's syndrome and that he had been experimenting with strobe photography to take pictures of other people with Tourette's. He could often, he said, catch his subjects in mid-tic. I very much liked his photos, and we decided to travel together, meeting his fellow Touretters around the world and documenting their lives and adaptations to this strange neurological condition.

The word "tic," in the context of Tourette's syndrome, covers a multitude of odd, repetitive, stereotyped, irrepressible behaviors. The simplest tics may consist of twitches or jerks, blinking, grimacing, shrugging, or sniffing. Other tics may be much more elaborate and complex. Lowell, for instance, fascinated by my old-fashioned pocket watch, developed an irresistible urge to tap it gently three times on the glass. (Once I teased him by moving the watch as he reached for it, then hiding it in a pocket. He became quite frantic at the frustration of his compulsion, and I had to produce the watch so he could satisfy his need.)

Most tics do not have any "meaning" to begin with but are

more akin to involuntary muscle (so-called myoclonic) jerks, though some tics may be elaborated or given meaning subsequently. Despite this, many of the tics and compulsions of Tourette's seem to be aimed at testing the boundaries of what is socially acceptable or, indeed, physically possible.

A person with Tourette's has a certain degree of voluntary control of an otherwise involuntary or compulsive behavior, so that, with a punching tic, for example, the fist will stop millimeters from someone's face. But Touretters may be less careful with themselves—I know two who have compulsions to fling themselves facedown onto the ground, and others who have broken bones or concussed themselves from violent blows to their own chests or heads.

Verbal tics, especially blurted-out obscenities or curses, are relatively rare in Tourette's, but they can cause deep offense—and here consciousness may step in to defuse the offending words. For example, Steve B., who feels compelled to shout "Nigger!," will at the last moment turn this into "Nickels and dimes!"

Tourettic behaviors are often at complete odds with the "real" person. Thus, when I first met Andy J., who has an irrepressible spitting tic, he struck the clipboard out of my hand and pointed to his wife, shouting, "She's a whore, and I'm a pimp"—but he is a sweet and even-tempered young man, with the most tender feelings for his wife.

And yet sometimes one feels that Tourette's may contribute a special creative energy. Samuel Johnson, the great eighteenth-century man of letters, almost certainly had Tourette's. He had many compulsions or rituals, especially upon entering a house, when he would twirl about or gesticulate in the doorway, then give a sudden spring, followed by a vast stride over the doorsill. He also exhibited strange vocalizations, litanic muttering, and involuntary mimicry of others. One cannot avoid thinking that his enormous spontaneity, antics, and lightning-quick wit had an organic connection with his accelerated, motor-impulsive state.

TOGETHER LOWELL AND I went to Toronto to visit Shane F., an artist who manages to make beautiful, compelling paintings and sculptures despite tics and compulsions so severe that his daily life is full of challenges and vicissitudes.

It was obvious at first glance that Shane had a different form of Tourette's than Lowell's. He was constantly in movement, constantly exploring. Everything and everyone around him would be looked at, palpated, turned over, prodded, scrutinized, smelled—a convulsive but at the same time playful investigation of the world around him. His senses seemed hyperacute; he noticed everything, and he could hear a whisper fifty yards away. He would run thirty or forty yards and then loop back—on the way, he might, with amazing agility, duck and run between someone's legs. And he had an anarchic sense of humor, often making multilayered, instant puns and jokes.

Shane has a particularly intense form of Tourette's, but he avoids the medications available to dampen down his tics and vocalizations. For him, they come at too great a price, since he feels they also dampen down his creativity.

One day the three of us strolled along a boulevard in Toronto—a sauntering broken by Shane's sudden dashes and occasional kneeling on the ground to smell or taste the asphalt. It was a perfect, sunny day, and we passed an open-air café, where at one sidewalk table, we saw a young woman bringing a delicious-looking hamburger to her mouth. Lowell and I felt our mouths watering, but Shane leapt into action and, with a lightning-quick lunge, took a large bite out of her hamburger before it reached her mouth.

The woman was stunned, as were her companions—but then she broke out into laughter. She saw the comedic aspect of Shane's bizarre act, and a potentially provocative episode was defused. There are not always such happy endings to Shane's sudden acts, which frequently go beyond the limits of social tolerance. Often he is regarded with suspicion; a number of times his unusual behavior has aroused the aggression of police or passersby. And his constant tics and compulsions can be exhausting to him and to those around him.

LOWELL AND I TRAVELED to Amsterdam, where we had been invited to appear together on a widely seen television show. I had fallen in love with Holland as a teenager—the place and, no less important, the sense of intellectual, moral, and creative freedom which had characterized its people since the days of Rembrandt and Spinoza. (When I first went to Holland, I was struck by the fact that the paper currency indicated its denominations in Braille as well as print.)

How, I wondered, would the Dutch see Tourette's? Would their freedom and independence of mind reduce the shock, fear, and anger that people with Tourette's can provoke?

The day before our television interview, we strolled around Amsterdam—I followed a few yards behind Lowell so that I could watch the reactions to his strange, sudden movements and noises. People's reactions appeared openly on their faces as they passed by us—some amused, some disturbed, and a few outraged.

Apparently, many people saw our television interview the following day, for when we went out again the next morning, the reactions were totally different. There were smiles, looks of curiosity, and friendly greetings, as people now seemed to recognize Lowell and to understand something about Tourette's. It drove home to us how essential it was to educate the public and transform their understanding—and how this could be done overnight, by a single television show.

That evening, relaxed, we went to a bar, where we were offered some pot and smoked it outside. We spent hours wandering around the city—looking at churches, at reflections in the canals, at shop windows, at people. Lowell, his camera at the ready, felt he was getting the best photos of his life. When we got back to our hotel later that night, and the bells of old churches started to toll, I had a sense of euphoria. Everything was right in the universe. This was the best of all possible worlds.

Lowell was less euphoric at breakfast the next morning, when he discovered that, in the happy confusion of being stoned, he had forgotten to load his camera with film, and the shots of a lifetime he thought he had captured were nonexistent.

In Rotterdam we met Ben van de Wetering, a brilliant Dutch psychiatrist who ran a clinic for patients with Tourette's—a rarity at the time. He introduced us to two of his patients. One was a rather Teutonic young man, very formal in dress and manner, who said that he detested his Tourette's and how it drew unwelcome attention to him. "It is completely *useless*!" he added, saying that he suppressed or converted his copralalia as much as possible. Thus whenever the word "Fuck!" was about to burst from his mouth, he was able to alter it, with effort, to "Frightful!" (This in fact attracted more attention than "fuck" would have done.) His Tourette's, in response to being suppressed or sanitized in daytime, took revenge on him at night, when a litany of obscenities escaped his sleeping lips.

The other patient, a young woman, was too shy or fearful to Tourette in public—but once she was "liberated" (as she put it) by Lowell's exuberant Tourette's, she allowed herself to Tourette along with him, in an amazing duet of convulsive movements and noises. She told me, "There is something primal in Tourette's—whatever I perceive or think or feel is instantly transformed into movements and sounds." She enjoyed this rushing stream; she felt it was "like life itself," but acknowledged that it could cause much trouble in social settings.

Tourette's, in its effects, is never confined to the person but spreads out and involves others and their reactions; and they, in turn, exert pressure—often disapproving, sometimes violent—on those with Tourette's. Tourette's cannot be studied or understood in isolation, as a "syndrome" confined to the person who has it; it invariably has social consequences and comes to include or incorporate these as well. What one sees, therefore, is a complex negotiation between the affected individual and his world, a form of adaptation sometimes humorous and benign, at other times charged with conflict, pain, anxiety, and rage.

. . .

THE FOLLOWING YEAR, Lowell and I took a road trip across the United States, visiting a dozen or so people with Tourette's who had agreed to meet with us.

Weaving through the outskirts of Phoenix with Lowell at the wheel was a remarkable experience, as he would suddenly stamp on the brakes or the accelerator, yanking the steering wheel to one side or the other. But once we were on the open road, Lowell's ticcy, impulsive, almost frenetic state gave way to one of stillness and concentration. He now sat calmly with his gaze fixed on the unfurling road ahead, which ran like an arrow through the central desert of Arizona. He kept our speed at an unvarying sixty-five miles per hour, never deviating.

At one point—we had been driving for three hours and needed to stretch our legs—I said to Lowell, "If you got out here and walked among the cacti, would you Tourette much?"

"No," he said. "What would be the point?"

Lowell has strong touching tics or compulsions; he cannot be around people without having to touch them. Usually he does this gently, with a hand or a foot. It is, one feels, almost an animal urge, the way a horse butts its head against a person and nuzzles them. People's reactions to being touched—whether positive, negative, or neutral—complete the circuit. But no such reaction could be expected from a plant.

This reminded me of a young Vietnamese man with Tourette's I once met. He had had some coprolalia in his native land, but now that he lived in San Francisco, where few people understood Vietnamese, he no longer uttered Vietnamese expletives. Like Lowell, he said, "What would be the point?"

. . .

SOMETIMES TOURETTERS ARE attracted by sudden bits of tactility or visual appearance—crumpledness, skewedness, odd asymmetries or shapes. (One Touretter, a wood-carver, likes to introduce sudden, convulsive asymmetries in his work—to make a chair "shaped like a tic or shriek.") Lowell often indulges in a compulsive repetition and permutation of odd words and sounds, whose very oddness provokes and gratifies the ear. At breakfast one morning, he got excited by the oatmeal, which he called "oakmeal," and kept repeating, "Oakmeal, oakmeal," then, after a while, an explosive "Kkkmmm!" At another meal, he seized upon the word "lobster," repeated "Lobbsster," "lobbsster," followed by "Mobbsster," "Slobbsster," finally concluding, "I love the sound and look of 'bbsstt.' "

"I take great pleasure in repeating words over and over again," he said. "It's the same feeling of satisfaction as I have with my compulsive touching—like having to touch the glass on your watch, feeling the click of my nail on the glass, reveling with different senses."

Tourette's can be exacerbated by hunger, and when we arrived in Tucson, having driven straight from Phoenix without a food stop, Lowell was racked by such violent tics that when we entered a restaurant, every eye was drawn to him. We sat at a table, and Lowell said, "I'm going to try something. Don't disturb me for fifteen minutes." He closed his eyes and started to breathe deeply and rhythmically, and within thirty seconds his tics were reduced; after a minute they were totally gone. When a waiter approached—he had observed Lowell's violent movements when we came in—I put a finger on my lips and waved him away. At exactly fifteen minutes, Lowell opened his eyes, looking very relaxed and almost tic-free. I could hardly

99

believe it—I would have thought such a change physiologically impossible.

"What happened? What did you do?" I asked Lowell. He said that he had learned Transcendental Meditation as a way of dealing with otherwise uncontrollable ticcing in public places. "It's just autohypnosis," he explained. "You have a mantra, a little word or phrase repeating slowly in your mind, and you soon get into a sort of trance and become oblivious to everything. It calms me down." He remained almost tic-free for the rest of the evening.[1]

Lowell had tracked down a pair of identical twins with Tourette's in Arizona. Tourette's had presented itself simultaneously in both boys—with sudden, screeching imitations of their pet cockatoo. Subsequently, they both developed shoulder shrugging, nose wrinkling, and clicking noises, followed by complex tics and contortions of the limbs and trunk. The picture was similar but not quite identical—one had winking tics, the other had gasping tics. And yet, unless one analyzed it minutely, the two of them looked and behaved in much the same way. How much of this was some sort of genetic predisposition, I wondered, and how much their tendency to imitate each other?

We met a young man in New Orleans with severe tics as well as obsessions and compulsions—a not uncommon combination. He had once worked in a missile silo in South Dakota, a job that terrified him because, having a compulsion to fiddle with switches, he was constantly afraid he would launch a missile and set off a nuclear war. The pay was good, and all his fel-

---

1. On another occasion we found ourselves in a shop filled with many clocks. Lowell felt alarmed when he saw all the pendulums swinging to and fro. "We can't stay here," he said. "I'll get hypnotized."

low workers friendly, but the ever-present sense of risk—though exciting—undid him, so he gave the job up for something less stressful.

In Atlanta we met Karla and Claudia, another pair of identical twins, who, like Shane, had that extravagant phantasmagoric form of Tourette's I sometimes think of as "super" Tourette's. They were fine, funny, intelligent young women in their early twenties, their voices hoarse from incessant shouting. They had many motor tics and contortions, too, but it was through their mouths that their often bizarre impulses and fantasies broke out.

Driving with Karla and Claudia was taxing—at each turn, one would shout "Right!" and the other "Left!" They told us how they had caused stampedes in movie houses by together shouting "Fire!" and cleared beaches by screaming "Shark!" They shouted with ear-splitting loudness from their bedroom window—"Black and white lesbians!" was one such cry, and another, more upsetting, was "My father is raping me!" Although all their neighbors knew about their wild Tourette shouting, their father could never get used to it and was cruelly distressed by the girls' shouts of "Rape!"

That our brief, patchy tour of America ended with such an extreme case is perhaps unfortunate, but such cases stay in the mind, and are often illuminating in their excess.

Lowell and I, traveling across America to meet a dozen Touretters and their families, had seen a wider range of Tourette's than one is ever likely to see in a hospital clinic, a much wider range than an ordinary neurologist would ever encounter. If there are extravagant forms, there are also forms so mild as to escape clinical notice—Tourette's, like autism, presents on a spectrum. One may have a very complex yet mild form of

Tourette's or a very simple yet severe form. And in any one person, Tourette's can fluctuate in intensity, as well as in form; there may be months or years of relative remission, and months or years of cruel exacerbation.

LOWELL HAD HEARD of an almost mythical place, in far northern Canada, where there is an entire community of Touretters, an extended Mennonite family that has had Tourette's among them for at least six generations (he had begun to refer to it as Tourettesville). What would it mean to be a member of this huge family in which ticcing and shouting is far from unusual, almost part of a family tradition? How might Tourette's affect or be affected by moral or religious beliefs, in such an isolated religious community? We decided to make a visit to find out.

At the airport closest to La Crete—it was little more than a landing strip in the woods—we rented a battered, splattered car, its windshield cracked from the coarse gravel on the roads. As we set off on the seventy-mile drive to La Crete, I felt the city tensions draining from me and observed Lowell's Tourettish outbreaks growing milder, soothed by the beauty, the peacefulness, the remoteness of the countryside. Reaching the village of La Crete, we passed a Mennonite couple selling watermelons by the road. We stopped and bought one, chatted a little: they had come from British Columbia, weaving from one tiny community to another, part of the quiet, half-religious, half-commercial network that binds the Mennonite communities of the Northwest together.

The Mennonites are descendants of a large group from Germany and the Low Countries that was driven to seek religious freedom first in the Ukraine and then in Canada. They still believe in a traditional farming way of life, in closeness to the soil and to family, in nonviolence, plainness, and a partial withdrawal from the great world outside.

In La Crete, a village of seven hundred people, there is a church for each of the five main Mennonite sects, and within the order there is a considerable range of practices and beliefs. The strictest are the Old Colony Mennonites, who are suspicious of the secular in education and daily life (but even they are not absolutely secluded like the Amish, a subgroup that split off in the 1690s). These conservative villagers dress soberly in dark clothes, the women wearing headdresses—but others in the town wear jeans and shirts. There is a simplicity and down-to-earthness about the place, along with a sense of tranquillity.

This tranquillity was broken when we reached David Janzen's house. David was the most pronounced Touretter here, and Lowell had arranged for us to meet him. Now David ran out to greet us, yelping and ticcing. The noises—ear shattering, shocking—seemed to disturb his whole being and, indeed, to disturb the whole placid face of La Crete. His cheery Tourettisms set off Lowell. They hugged each other, ticced, yelped—it was both affecting and absurd, reminding me of the excitement of two dogs meeting.

David, now in his early forties, had started having varied tics when he was eight. They were no surprise, because his mother and two older sisters also had them, as well as dozens of cousins and more distant family members. They were called "the fidgets," and it was said that the Janzens were "restless" or "nervous."

"Grandmother would always be blinking her eyes and smacking her lips," said a cousin of David's, "or clucking or hooting or making faces, whatever—it was just normal. Everybody did it."

David's real struggle started when he was fifteen, when he started to shout "Fuck!" in a loud voice. Such uttering of obscenities and profanities was not a common manifestation of Tourette's in La Crete. Unlike the fidgets, the cursing smacked of a savage self or a vile prompting from the Devil. David's compulsions started to multiply. Sometimes he had the impulse to hurt himself or break things. "Devil!" he would say to himself. "Why don't you get out of me and leave me alone?"

The boy fled, but inward. "When I had the spell of the curse, I more or less stayed home," he said. "I didn't communicate with people—maybe for a year. Often at that time I just went to my room and cried myself to sleep."

David's parents tried to be understanding, but they, too, were confused. They saw his peculiar illness as half moral, half physical. They felt that David was subject to some outer force but also that he "allowed" the cursing. He, too, began to see himself as weak-willed. Some in La Crete had a simpler view: David was the object of divine anger and punishment. According to one villager, there was the feeling at the time that "the Janzens were strange, especially Davy. God must be punishing this family for something."

In his early twenties, David married and started a family, though his disturbances continued. Frequently he would feel compelled to pant violently or hold his breath; these respiratory convulsions, which are not uncommon in Tourette's, were exhausting. "I was getting so tired because I was fighting it so hard, especially if I would drive," he recalled. He used to drive a truck from High Level to Hay River, fighting compulsions to suddenly brake or accelerate or swerve. Sometimes David would injure himself with ticcish movements. "I was once operating a chain saw and cut my leg—I know now that the movement was a Tourette," he said, showing me the long white scar on his left knee.

David loved the hard work of farming, loved working with cattle and horses, but had difficulty because his tics startled the animals and caused them to shy. He had to give up work by the time he was thirty, but on welfare his morale sank lower and lower. Finally, at thirty-eight, David reached a crisis. "I felt I had to have an answer, answers, or else I couldn't go on."

A local doctor had told him that he might have Huntington's chorea, an appalling and fatal disease. In Edmonton he was told it might be myoclonus, which causes the muscles to contract suddenly. Finally David was sent to Dr. Roger Kurlan, a neurologist specializing in movement disorders at the University of Rochester in New York.

Kurlan took one look at David and said, "You've got Tourette's syndrome." David had never heard of it. As Kurlan described the tics, the compulsions, David felt immense relief. "It made me want to jump for joy," he said. "It took away the terrible feeling of a curse. It was not the Devil working in me—which was my worst fear—and it was not medical doom. I had a simple disease, and it even had a name. A pretty name too—I kept on repeating it."

But one point puzzled David. "Unusual, you say," he repeated to Kurlan. "Doesn't it run in families?"

"I rarely see it in families," said the doctor.

"Well," said David, a bit surprised, "most everyone I know has got Tourette's. My family, anyhow—my mother, my two sisters." And pulling out a pencil, he drew a family tree on the blotter, indicating more than a dozen close family members who were also affected.

When I spoke to Kurlan four years later, he told me that this was the most astonishing moment of his entire medical life. He had never considered that the syndrome could have such a strong genetic component. He made a visit to La Crete, still incredulous. He explored the village day and night for a week, interviewing a total of sixty-nine Janzen family members. Kurlan told them that what they had was neither a grave organic disease nor a curse, but a nonprogressive and probably genetically determined disorder of the nervous system.

This scientific explanation, though cause for much relief and discussion, did not entirely displace the religious view. Behind Tourette's, the people of La Crete still saw the hand of God. But they fully embraced the term: odd behavior in La Crete is now called Touretting. Georges Gilles de la Tourette, the nineteenth-century French neurologist who identified the syndrome, would be astonished to find his name known—indeed, common currency—in a remote farming village four thousand miles from Paris.

There is, among Orthodox Jews, a blessing to be said on witnessing the strange: one blesses God for the diversity of his creation, and one gives thanks for the wonder of the strange. This, it seemed to me, was the attitude of the people of La Crete to the Tourette's in their midst. They accepted it not as something annoying or insignificant, to be reacted to or overlooked, but as a deep strangeness, a wonder, an example of the absolute mysteriousness of Providence.

Touretters, with their impulses and cursings, can feel outcast, singled out by an unusual condition that no one around them shares or fully understands. Many have found themselves avoided or punished as children and barred as adults from restaurants and other public places. Lowell had faced this for years, and for him, therefore, La Crete was particularly sweet—it provided the first exemption he had ever known from a negative attention. Part of him fell in love with La Crete, so that he had visions of marrying a nice Mennonite girl with Tourette's and living there happily ever after. "I felt the lure of New York," Lowell reflected after we left, "but I also felt the lure of spending a life with a family and friends in a place like Tourettesville. But I was just a visitor, a very loved visitor, but still just a visitor. I was only part of their world for a very short time."

# Urge

Walter B., an affable, outgoing man of forty-nine, came to see me in 2006. As a teenager, following a head injury, he had developed epileptic seizures; these first took the form of attacks of déjà vu that might occur dozens of times a day. Sometimes he would hear music that no one else could hear. He had no idea what was happening to him and, fearing ridicule or worse, kept his strange experiences to himself.

Finally he consulted a physician, who made a diagnosis of temporal lobe epilepsy and started him on a succession of anti-seizure drugs. But his seizures—both grand mal and temporal lobe—became more frequent. After a decade of trying different anti-seizure drugs, Walter consulted another neurologist, an expert in the treatment of "intractable" epilepsy, who suggested a more radical approach: surgery to remove the seizure focus in his right temporal lobe. This helped a little, though a few years later, a second, more extensive operation was needed. The second surgery, along with medication, controlled his seizures more effectively but almost immediately led to some singular problems.

Walter, previously a moderate eater, developed a ravenous appetite. "He started to gain weight," his wife later told me, "and his pants changed three sizes in six months. His appetite was out of control. He would get up in the middle of the night and eat an entire bag of cookies, or a block of cheese with a large box of crackers."

"I ate everything in sight," Walter said. "If you put a car on the table, I would have eaten it." He became very irritable, too, he told me:

> I raged for hours at inappropriate things at home (no socks, no rye bread, perceived criticisms). Driving home from work a driver squeezed me on a merge. I accelerated and cut him off. I rolled my window down, gave him the finger, and began screaming at him, and threw a metal coffee mug and hit his car. He called the police from his cell. I was pulled over and ticketed.

Walter's attention assumed an all-or-none quality. "I became distracted so easily," he said, "that I couldn't get anything started or done." Yet he was also prone to getting "stuck" in various activities—playing the piano, for example, for eight or nine hours at a time.

Even more disquieting was the development of an insatiable sexual appetite. "He wanted to have sex all the time," his wife said.

> He went from being a very compassionate and warm partner to just going through the motions. He didn't remember having just been intimate. . . . He wanted sex constantly after surgery . . . at least five or six times a day. He also gave up on foreplay. He would always want to get right to it.

There were only fleeting moments of satiety, and within seconds of orgasm, he wanted intercourse again and again. When his wife became exhausted, he turned to other outlets. Walter had always been a devoted and thoughtful husband, but now his sexual desires, his urges, spread beyond the monogamous heterosexual relationship he had enjoyed with his wife.

It was morally inconceivable for him to force his sexual attentions on a man, woman, or child—internet pornography, he felt, was the least harmful answer; it could provide some sort of release and satisfaction, even if only in fantasy. He spent hours masturbating in front of his computer screen while his wife slept.

After he started viewing adult pornography, various websites solicited him to purchase and download child pornography, and he did. He became curious, too, about other forms of sexual stimulation—with men, with animals, with fetishes.[1] Alarmed and ashamed of these new compulsions, so alien to his previous sexual nature, Walter found himself in a grim struggle for control. He continued to go to work, to go out socially, to meet his friends for meals or movies. During these times he was able to keep his compulsions in check, but at night, alone, he gave in to his urges. Deeply ashamed, he told no one of his predicament, living a double life for more than nine years.

Then the inevitable happened, and federal agents came to Walter's house to arrest him for possession of child pornography. This was terrifying, but it was also a relief, because he no longer had to hide or dissimulate—he called it "coming out of the shad-

1. Such "polymorphous perversion" (as Freud called it) may occur in a number of conditions where dopamine levels in the brain are too high. It developed in some of my postencephalitic patients "awakened" by L-dopa, and it can occur in association with Tourette's syndrome or chronic use of amphetamines or cocaine.

ows." His secret was now exposed to his wife and his children, and to his physicians, who immediately put him on a combination of drugs that diminished—indeed, virtually abolished—his sexual drive, so that he went from an insatiable libido to almost no libido at all. His wife told me that his behavior instantly "reverted back to loving and compassionate." It was, she said, as if "a faulty switch was turned off"—a switch that had no middle position between on and off.

I saw Walter on several occasions in the time between his arrest and his prosecution, and he expressed fear—mostly of the reactions of his friends, colleagues, and neighbors. ("I thought they would point fingers or throw eggs at me.") But he thought it unlikely that a court would view his conduct as criminal, in view of his neurological condition.

On this point, Walter was wrong. Fifteen months after his arrest, his case finally came to court, and he was prosecuted for downloading child pornography. The prosecutor insisted that his so-called neurological condition was of no relevance, a red herring. Walter, he argued, was a lifelong pervert, a menace to the public, and should be put away for the maximum term of twenty years.

The neurologist who had originally suggested temporal lobe surgery and had treated Walter for almost twenty years appeared in court as an expert witness, and I submitted a letter to be read in court, explaining the effects of his brain surgery. We both pointed out that Walter's condition was a rare but well-recognized one called Klüver-Bucy syndrome, which manifests itself as insatiable eating and sexual drive, sometimes combined with irritability and distractibility, all on a purely physiological basis. (The syndrome had first been recognized in the 1880s, in lobectomized monkeys, and subsequently described in human beings.)

The all-or-none reactions that Walter had shown were characteristic of impaired central control systems; they may occur, for example, in parkinsonian patients on L-dopa.[2] Normal control systems have a middle ground and respond in a modulated fashion, but Walter's appetitive systems were continually on "go"—there was scarcely any sense of consummation, only the drive for more and more. Once his physicians became aware of the problem, medication readily brought it under control—albeit at the cost of a sort of chemical castration.

In court, his neurologist emphasized that Walter was no longer subject to his sexual urges and pointed out that he had never actually laid hands on anyone other than his wife. (He also noted that, among more than thirty-five cases on record of pedophilia associated with neurological disorders, only two had been arrested and charged with criminal behavior.) In my own letter to the court, I wrote:

> Mr. B. is a man of superior intelligence and . . . moral sensibility, who at one point was driven to act out of character under the spur of an irresistible physiological compulsion. . . . He is strictly monogamous. . . . There is nothing in his history

2. This also happened with many of my *Awakenings* patients, who had damage to various drive systems in their brains. Thus Leonard L. was, as he later said, a "castrate" with no libido at all before he received L-dopa, but on L-dopa, he developed a ravenous sexual appetite. He suggested that the hospital make a brothel service available for L-dopa-charged patients, and when his plans were frustrated, he masturbated constantly, and often openly, for hours.

or his current ideation to suggest that [he] is a pedophile. He poses no risk to children or to anyone else.

At the end of the trial, the judge agreed that Walter could not be held accountable for having Klüver-Bucy syndrome. But he *was* culpable, she said, for not speaking sooner about the problem to his doctors, who could have helped, and for persisting for many years in behavior that was injurious to others. She emphasized that his crime was not a victimless one.

She sentenced him to twenty-six months in prison, followed by twenty-five months of home confinement and then a further five-year period of supervision. Walter accepted his sentence with a remarkable degree of equanimity. He managed to survive prison life with relatively little trauma and made good use of his time in jail, establishing a musical band with some fellow inmates, reading voraciously, and writing long letters (he often wrote to me about the neuroscience books he was reading).

His seizures and his Klüver-Bucy syndrome remained well controlled by medication, and his wife stood by him throughout his years of prison and home confinement. Now that he is a free man, they have largely resumed their previous lives. They still go to the church where they were married many years ago, and he is active in his community.

When I saw him recently, he was clearly enjoying life, relieved that he had no more secrets to hide. He radiated an ease I had never seen in him before.

"I'm in a real good place," he said.

# The Catastrophe

In July of 2003, my neurological colleague Orrin Devinsky and I were consulted by Spalding Gray, the actor and writer who was famous for his brilliant autobiographical monologues, an art form he had virtually invented. He and his wife, Kathie Russo, had contacted us in regard to a complex situation that had developed after Spalding suffered a head injury, two summers earlier.

In June 2001, they had been vacationing in Ireland to celebrate Spalding's sixtieth birthday. One night while they were driving on a country road, their car was hit head-on by a veterinarian's van. Kathie was at the wheel; Spalding was in the back seat with another passenger. He was not wearing a seat belt, and his head crashed against the back of Kathie's head. Both were knocked unconscious. (Kathie suffered some burns and bruises but no permanent harm.) When Spalding recovered consciousness, he was lying on the ground beside their wrecked car, in great pain from a broken right hip. He was taken to the local rural hospital and then, several days later, to a larger hospital, where his hip was pinned.

His face was bruised and swollen, but the doctors focused on his hip fracture. It was not until another week went by and the swelling subsided that Kathie noticed a "dent" just above Spalding's right eye. At this point, X-rays showed a compound fracture of the eye socket and skull, and surgery was recommended.

Spalding and Kathie returned to New York for the surgery, and MRIs showed bone fragments pressing against his right frontal lobe, though his surgeons did not see any gross damage to this area. They removed the fragments, replaced part of his skull with titanium plates, and inserted a shunt to drain away excess fluid.

He was still in some pain from his hip fracture and could no longer walk normally, even with a braced foot (his sciatic nerve had been injured in the accident). Yet, strangely enough, during these terrible months of surgery, immobility, and pain, Spalding seemed in surprisingly good spirits—indeed, his wife thought he was "incredibly well" and upbeat.

Over Labor Day weekend of 2001, five weeks after his brain surgery and still on crutches, Spalding gave two performances to huge audiences in Seattle. He was in excellent form.

Then, a week later, there was a sudden, profound change in his mental state, and Spalding fell into a deep, even psychotic, depression.

NOW, TWO YEARS AFTER the accident, on his first visit to us, Spalding entered the consulting room slowly, carefully lifting his braced right foot. Once he was seated, I was struck by his lack of spontaneous movement or speech, his immobility and lack of facial expression. He did not initiate any conversation, and he responded to my questions with very brief, often single-word, answers. My first thought, and Orrin's, was that this was not simply depression, or even a reaction to the stress and the surgeries of the past two years—to my eye, it clearly looked as if Spalding had neurological problems as well.

When I encouraged him to tell me his story in his own way, he began—rather strangely, I thought—by relating how, a few months before the accident, he had had a sudden "compulsion" to sell his house in Sag Harbor, which he loved and in which he and his family had lived for five years. He and Kathie agreed that the family needed more room, so they bought a house nearby, with more bedrooms and a bigger yard. Nonetheless, Spalding had resisted selling the old house, and they were still living in it when they left for Ireland.

It was while he was in the hospital in Ireland following his hip surgery, he told me, that he finalized a deal to sell the old house. He later came to feel that he was "not himself" at the time, that "witches, ghosts, and voodoo" had "commanded" him to do it.

Even so, despite the accident and the surgeries, Spalding remained in high spirits during the summer of 2001. He felt full of new ideas for his work—the accident, even the surgeries, would be wonderful material—and he could present them in a new performance piece, entitled *Life Interrupted*.

I was struck, and perhaps disquieted a little, by the readiness with which Spalding was prepared to turn the horrifying events of the summer to creative use. Yet I could also understand it, because I had not hesitated, in the past, to use some of my own crises as material in my books.

Indeed, using one's own life (and sometimes others' lives) as material is common among artists—and Spalding was a very special sort of artist. Although he acted in television and films from time to time, his true originality was expressed in the dozen or so highly acclaimed monologues that he performed onstage. (A number of these, such as *Swimming to Cambodia* and *Monster in a Box,* were filmed.) His stagecraft was stark and simple: alone on a stage, with nothing but a desk, a glass of water, a notebook, and a microphone, he would establish an immediate rapport with the audience, spinning webs of largely autobiographical stories. In these performances, the comedies and mishaps of his life—the often absurd situations he found himself in—were raised to an extraordinary dramatic and narrative intensity. When I inquired about this, Spalding told me that he was a "born" actor—that, in a sense, his whole life was "acting." He wondered sometimes if he did not create crises just for material—an ambiguity that worried him. Had he sold his house as "material"?

One of the special features of Spalding's monologues was that, onstage at least, he rarely repeated himself; the stories always came out in slightly different ways, with different emphases. He was a gifted inventor of the truth, of whatever seemed true to him at the moment.

THE FAMILY WAS DUE to move out of the old house on September 11, 2001. By then, Spalding was already consumed with regret over selling it, a decision he regarded as "catastrophic." When Kathie told him about the attack on the World Trade Center that morning, he barely registered it.

Ever since, Kathie said, Spalding had been sunk in depressive, obsessive, angry, guilty rumination about selling the house.

Nothing could distract him from it. Scenes and conversations about the house replayed incessantly in his mind. All other matters seemed to him peripheral and insignificant. Previously a voracious reader and a prolific writer, he now felt unable to read or write.

Spalding had had occasional depressions, he said, for more than twenty years, and some of his physicians thought that he had a bipolar disorder. But these depressions, though severe, had yielded to talk therapy or, sometimes, to treatment with lithium. His current state, he felt, was different. It had unprecedented depth and tenacity. He had to make a supreme effort of will to do things like ride his bicycle, which he had previously done spontaneously and with pleasure. He tried to converse with others, especially his children, but found it difficult. His ten-year-old son and his sixteen-year-old stepdaughter were distressed, feeling that their father had been "transformed" and was "no longer himself."

In June of 2002, Spalding sought help at Silver Hill, a psychiatric hospital in Connecticut, where he was put on Depakote, a drug sometimes used for bipolar disorder, but there was little improvement in his condition, and he became more and more convinced that some sort of irresistible, evil Fate had drawn him in and commanded him to sell the house.

In September 2002, Spalding jumped off his sailboat into the harbor, planning to drown himself (he lost his nerve and clung to the boat). A few days later, he was found pacing on the Sag Harbor bridge, eyeing the water, until the police intervened and Kathie took him home.

Soon after this, Spalding was admitted to the Payne Whitney Psychiatric Clinic, on the Upper East Side. He spent four months there and was given more than twenty shock treatments and drugs of all kinds. He responded to none of them and, indeed, seemed to be getting worse by the day. When he emerged from Payne Whitney, his friends felt that something terrible and perhaps irreversible had happened. Kathie thought that he was "a broken man."

In June of 2003, hoping to clarify the nature of his deterioration, Spalding and Kathie went to UCLA's Resnick Hospital for neuropsychiatric testing. He did badly on various tests, which showed "attentional and executive deficits typical of right frontal lobe damage." The doctors there told him that he might deteriorate further, because of cerebral scarring where the frontal lobe had borne the impact of the crash and the imploded bone fragments. They told him that he might never be capable of original work again. According to Kathie, Spalding was "morally devastated" by their words.


IN JULY, when Spalding first came to see Orrin and me, I asked him if there were any other themes besides the sale of his house that he ruminated about. He said yes: he often thought about his mother and the first twenty-six years of his life. It was when he was twenty-six that his mother, who had been intermittently psychotic since he was ten, fell into a self-torturing, remorseful state, focused on the selling of her family house. Unable to endure her torment, she had committed suicide.

In an uncanny way, he said, he felt that he was recapitulating what had happened with his mother. He felt the attraction of suicide and thought of it constantly. He said he regretted not having committed suicide at the UCLA hospital. Why there? I inquired. Because one day, he replied, someone had left a large plastic bag in his room—and it would have been "easy." But he was pulled back by the thought of his wife and children. Nevertheless, he said, the idea of suicide rose "like a black sun" every day. He said the past two years had been "gruesome" and added, "I haven't smiled since that day."

Now, with his partly paralyzed foot and the brace, which irritated him if used for any length of time, he was also denied physical outlets. "Hiking, skiing, and dancing had been a huge factor in my mental stability," he told me, and he felt, too, that he had been disfigured by the injury and by the surgery to his face.

THERE WAS A BRIEF, dramatic break in Spalding's rumination just a week before he came to see us, when he had to have surgery because one of the titanium plates in his skull had shifted. The operation took four hours, under general anesthesia. Coming to from the anesthesia and for about twelve hours afterward, Spalding was his old self, talkative and full of ideas. His rumination and hopelessness had vanished—or, rather, he now saw how he could use the events of the past two years creatively in one of his monologues. But by the next day this brief excitement or release had passed.

As Orrin and I talked over Spalding's story and observed his peculiar immobility and lack of initiative, we wondered whether an organic component, caused by the damage to his frontal lobes, had played a part in his strange "normalization" after anesthesia. It seemed as if his compromised frontal lobes no longer allowed him any middle ground, either paralyzing him in an iron neurological restraint or suddenly, briefly, releasing him into an opposite state. Had some sort of buffer—a protective, inhibiting frontal-lobe function—been breached by his accident, allowing an uncontrollable rush of previously suppressed or repressed thoughts and fantasies into his consciousness?

The frontal lobes are among the most complex and recently evolved parts of the human brain—they have vastly enlarged over the past two million years. Our power to think spaciously and reflectively, to bring to mind and hold many ideas and facts, to attend to and maintain a steady focus, to make plans and put them into motion—these are all made possible by the frontal lobes.

But the frontal lobes also exert an inhibiting or constraining influence on what Pavlov called "the blind force of the subcortex"—the urges and passions that might overwhelm us if left unchecked. (Apes and monkeys, like children, though clearly intelligent and capable of forethought and planning, have less developed frontal lobes and tend to do the first thing that occurs to them, rather than pausing to reflect. Such impulsivity also can be striking in patients with frontal-lobe damage.) There is normally a beautiful balance, a delicate mutuality, between the frontal lobes and the subcortical parts of the brain that mediate perception and feeling, and this allows a consciousness that is free-ranging, playful, and creative. The loss of this balance through frontal-lobe damage can "release" impulsive behaviors, obsessive ideas, and overwhelming feelings and compulsions. Were Spalding's symptoms a result of frontal-lobe damage or severe depression, or a malignant coupling of the two?

Frontal-lobe damage can lead to difficulties with attention and problem-solving and impoverishment of creativity and intellectual activity. Although Spalding felt that he had not had any intellectual deterioration since the accident, Kathie wondered whether his unceasing rumination might not, in part, be a "cover" or "disguise" for an intellectual loss that he did not want to admit. Whatever the case, Spalding felt that he could no longer achieve the high creative level, the playfulness and mastery, of his pre-accident performances—and others felt this, too.

I SAW SPALDING AGAIN in September 2003, with Kathie, two months after our initial consultation. He had been living at home, feeling very grim, unable to work. When I asked whether he felt any different, he said, "No difference." When I remarked that he appeared more animated and less agitated, he said, "People say so. I don't feel it." And then (as if to disabuse me of any notion that he might be better) he told me that he had staged a suicide "rehearsal" during the previous weekend. Kathie was away at a business conference in California, and, fearing for his safety in the country, she had arranged for him to spend the weekend in their Manhattan apartment. Nevertheless, he told me, he had set out for an excursion on Saturday with an eye to casing the Brooklyn Bridge and the Staten Island Ferry as suitable venues for a dramatic suicide, but he was "just too afraid" to act—particularly when he thought of his wife and children.

He had resumed cycling a little and often rode past his former home, though he could hardly bear to see it repainted, in the possession of others. He had offered to buy it back, thinking that this might release him from the "evil spell" cast on him, but its new owners were not interested.

Yet, Kathie pointed out, despite being deeply depressed and obsessed, Spalding had pushed himself during the past two years to travel and give several performances in other cities. But these shows, in which he recounted the accident, were far from his best. At one theater, he knocked on the stage door before the performance, and the director, who knew him well, took him at first to be a homeless man—he looked disheveled and unkempt. Spalding seemed distracted while he was onstage there and alienated the audience.

As we concluded our appointment, Kathie added that Spalding was due to go into the hospital the next day for an attempt to free his right sciatic nerve from the scar tissue that embedded it. His surgeon hoped the procedure might permit some regeneration of the nerve and allow him to move his foot properly. He would be having general anesthesia, and, remembering how anesthesia had affected him so dramatically a couple of months earlier, I arranged to visit him in the hospital a few hours after the operation.

When I arrived, I found Spalding remarkably animated and sociable, with a spontaneity I had not previously seen in him—a picture very unlike that of the almost mute, unresponsive man who had come to my office the day before. He started a conversation, offered me a cup of tea, inquired where I had traveled from, and asked what I was writing. He said that his obsessive rumination had totally ceased for two or three hours after the anesthesia wore off, and was still much reduced.

I visited again the next day—it was September 11, 2003, two years since he had fallen into his "evil" depression. He continued to be animated and conversational. Orrin, on a separate visit, was also able to have a "normal conversation" with Spalding. We were both amazed at this almost instantaneous reversal.

Orrin and I again speculated as to what might have allowed

this temporary "normalization." Orrin felt that, for nearly forty-eight hours, the anesthesia had damped down or inhibited the rumination and the negative feelings that Spalding's frontal-lobe damage had released; the anesthesia, in effect, provided the protective barrier that intact frontal lobes would normally provide.

On a third visit, early in the morning of September 12th, I again found Spalding in a good mood. He said that he had very little postoperative pain, and he got out of bed with alacrity to show how well he could walk without either crutches or a splint (though there was no neurological recovery as yet, and he had to lift the impaired foot high as he walked). As I was leaving, he asked me where I was going—the kind of friendly question he had scarcely asked in his self-involved state. When I said I was going swimming, he said that he, too, had a passion for swimming, especially in a lake near his house, and that he hoped to swim there when he got out of the hospital.

I was happy to observe a notebook on his table. (He had told me that he kept a journal while in the hospital in Ireland.) I said I thought that two years of torment was enough: "You have paid your dues to the powers of darkness." Spalding half-smiled and said, "I think so, too."

I felt guardedly optimistic at this point. Perhaps he was emerging, finally, from both his depression and his frontal-lobe injury. I told Spalding that I had seen many patients with more severe head injuries who, with time and the brain's ability to compensate for damage, had regained most of their intellectual powers.

· · ·

I HAD PLANNED to visit Spalding again the next day, but I was diverted by a phone message from Kathie saying that he had left the hospital without checking out, and without any money or identification.

The following morning, I found another message, this one telling me that Spalding had made his way to the Staten Island Ferry and then left a phone message saying that he was contemplating suicide. Kathie had called the police, who finally picked him up around ten p.m.—he had been riding back and forth on the ferry. He was admitted as an involuntary patient to a hospital on Staten Island, then transferred to a special brain-rehab unit at the Kessler Institute, in New Jersey, where Orrin and I saw him a few days later.

Spalding was very conversational and showed me fifteen pages he had just written—his first writing in many months. But he still had some strange and ominous obsessions—one had to do with what he called "creative suicide." He regretted that, having spoken to a reporter who was working on a magazine article about him, he had not taken her on the Staten Island Ferry and demonstrated a creative suicide there and then. I was at pains to say that he could be much more creative alive than dead.


SPALDING RETURNED HOME, and when I saw him on October 28th I was pleased to hear that he had performed two monologues in the past couple of weeks. When I asked how he could manage this, he emphasized a sense of commitment: if he had agreed to do something, he would do it, however he felt. Perhaps, too, he hoped that these performances would reenergize him. In the old days, Kathie told me, he would remain energized after a show and entertain friends and fans backstage. Now, although he would become somewhat animated in the act of performing, he would fall back into his depression almost as soon as the show was over.

After one of these performances, he left Kathie a note saying that he was going to jump off a bridge on Long Island—and he did jump. He felt he could not go back on this "commitment." This was a very public jump—he was observed by a number of witnesses, one of whom helped him back to shore.

Spalding wrote frequent suicide notes, which Kathie or the children would find on the kitchen table; the family would be thrown into a state of intense anxiety until he reappeared.

In November, Orrin and I went to see one of Spalding's performances; we were impressed by his professionalism and his virtuosity onstage but felt that he was still submerged in his memories and fantasies, not mastering and transforming them as he had once done.

SPALDING AND KATHIE CAME to see me again in early December. When I went to usher them into my office, Spalding's eyes were closed, and he seemed to be asleep—but he opened them at once when I spoke to him, and he followed me into the consulting room. He had not been asleep, he indicated, but "thinking."

"I still have enormous problems with rumination," he said. "I feel destined to follow my mother in a sort of self-hypnosis. It's all over, terminal. I'd be better off dead. What do I have to give?"

A week earlier, Spalding and Kathie had taken a boat trip, and she became frightened by the "purposeful" way he eyed the water—she felt she had to watch him all the time now.

When I told Spalding how impressed people were by his latest monologues, he said, "Yes, but that's because they see the old me, the way I was, even though that's gone. They're just sentimental and nostalgic."

I asked him whether transforming the events of his life, especially some of the very negative events, into a monologue enabled him to integrate them, and thus defuse them. He said no, not now. He felt that his current monologues, far from helping him as they once would have, merely aggravated his melancholic thinking. "Previously," he added, "I was on top of the material; I had the use of irony."

He spoke of being "a failed suicide" and asked me, "What would you do if your only choice was between institutionalization in a mental hospital and suicide?"

He said that his mind was filled with fantasies of his mother and of water, always water. All his suicidal fantasies, he said, related to drowning.

Why water, why drowning? I asked.

"Returning to the sea, our mother," he said.

This reminded me of the Ibsen play *The Lady from the Sea*. I had not read it for thirty years, but now I reread it—Spalding, a playwright himself, had surely read it—and was reminded how Ellida, who grew up in a lighthouse, the daughter of an insane mother, was herself driven to a sort of insanity by her obsession with the sea and what she felt as a "terrifying attraction" to a sailor who seemed to embody the sea. ("All the force of the sea is in this man.")

Moving to another house, for Ellida, as for Spalding, played a part in tipping her into a near-psychotic state, in which quasi-hallucinatory images of the past and of what she felt to be her "destiny" surged up like the sea from her unconscious, almost drowning her ability to live in the present. Wangel, her physician husband, saw the power of this: "This hunger for the boundless, the infinite—the unattainable—will finally drive your mind out completely into darkness." This was my fear now for Spalding—that he was being drawn towards death by powers that neither he nor I nor any of us could contend with.

Spalding had spent more than thirty years on "the slippery slope," as he called it, as a high-wire performer, a funambulist, and had never fallen off. He doubted if he could continue. While I expressed hope and optimism outwardly, I now shared his doubt.

ON JANUARY 10, 2004, Spalding took his children to a movie. It was Tim Burton's *Big Fish,* in which a dying father passes his fantastical stories on to his son before returning to the river, where he dies—and perhaps is reincarnated as his true self, a fish, making one of his tall tales come true.

That evening, Spalding left home, saying he was going to meet a friend. He did not leave a suicide note, as he had so often before. When inquiries were made, one man said he had seen him board the Staten Island Ferry.

Two months later, Spalding's body was washed up by the East River. He had always wanted his suicide to be high drama, but in the end he said nothing to anyone; he simply disappeared from sight and silently returned to the sea, his mother.

# Dangerously Well

M r. K. was an intelligent and cultivated seventy-two-year-old man, successful in the fashion industry and generally in good health. In September 2000, though, two years before he first came to see me, Mr. K. had complained of joint pains, and his physician had diagnosed him as having polymyalgia rheumatica and put him on prednisone, ten milligrams twice daily. Within days, the pain and stiffness subsided and Mr. K. experienced a great sense of well-being—too great, perhaps. The steroids, he later told me, "made me feel tremendously energetic. I had been walking like a ninety-year-old man, but now when I walked I felt as if I was flying. I never felt better in my life." His "euphoria" (as he was to call it in retrospect) increased over the ensuing months; he became more sociable and bolder in business dealings. He seemed, to himself and to those around him, to be in exuberantly good spirits.

It was not apparent that there was anything amiss until March of 2001, when he went to Paris on a business trip. There were hints of disorganization and excitement while he was preparing for the trip, and in Paris, these symptoms became full-blown: he forgot important appointments (which brought matters to his family's attention), bought more than a hundred thousand dollars' worth of art books, had altercations with the hotel staff, and assaulted a policeman at the Louvre.

This precipitated his admission to a French psychiatric hospital, where he appeared "grandiose and disinhibited" and confessed that, unknown to anyone else, he had increased his prednisone dosage to five times his original prescription. He had been at this elevated dose for at least three months. Clearly this high dosage had produced what is known as a "steroid psychosis," and Mr. K. was diagnosed as having "a manic episode with psychotic features." He was put on tranquilizers for the mania, and his prednisone dose was reduced to the original ten milligrams twice a day. But this had little effect, and after some days in the French hospital, still noisy and disinhibited, he was returned to New York on April 30, 2001, in the company of a physician.

Back in New York, Mr. K. was again admitted to a psychiatric ward; despite the drastic reduction in his steroids, he still seemed psychotic and markedly disorganized in his thinking. Neuropsychological tests showed a decline in his previously superior IQ, memory, language, and visuospatial functions.

Since no evidence could be found for any infectious, inflammatory, or toxic-metabolic cause for Mr. K.'s persisting cognitive deficits, his physicians felt that there must be a rapidly progressing neurodegenerative disease—in addition to (and perhaps predisposing him to or unleashed by) the steroid psychosis. Alzheimer's disease, Lewy body disease, and especially frontotemporal dementia were considered.

MRI and PET scan imaging of Mr. K.'s brain revealed reduced metabolism bilaterally—an inconclusive finding, but one that, together with his neuropsychological tests, was compatible with early dementia.

When Mr. K. was finally discharged home in early June, after six weeks in the hospital, he became more agitated and confused than ever, on one occasion attacking his wife. He now needed round-the-clock supervision, and he was admitted to a locked Alzheimer's facility. Here things rapidly went from bad to worse. He started to hoard food, to steal belongings from other residents, to become dirty and dilapidated—a drastic change for this previously meticulous dresser.

His wife, dismayed by the rapid disintegration of her husband, sought a new neurological opinion in mid-July. Mr. K.'s new doctor ordered more tests and began to taper Mr. K.'s dosage of prednisone further.

By September of 2001, after a year of continuous administration, Mr. K.'s steroids were finally stopped completely. His confusion diminished almost immediately. This was strikingly clear at a family wedding in the middle of the month—Mr. K., restored to his former dapper appearance, recognized most of the guests, greeting and chatting with them in a way that would have been inconceivable a month before, when he had been so grossly demented.

Mr. K. had already returned to his business, and neuropsychological tests a couple of weeks later showed great improvement in almost all of his cognitive functions, although there were still hints of impulsiveness, perseveration, and some intellectual deficits.

All this was very reassuring, but perplexing, too, for diseases like Alzheimer's and frontotemporal dementia are progressive—they do not go away virtually overnight. And yet here was Mr. K., at one point expected to spend the rest of his days in a locked Alzheimer's facility, now restored to his family, his business, to life as usual—as if suddenly awakened from a hideous, months-long nightmare. (His wife, who wrote a narrative of her husband's experiences, titled it "A Journey to Hell and Back.")

I FIRST MET Mr. K. about six months after this, in March 2002. He was a tall, amiable man, well dressed, affable, and voluble. He presented his story rationally and consecutively, but with numerous asides. (It was not clear how much he was recollecting his own experiences, and how much he had been told his story by others, and now had it well rehearsed and fluent.) He was persuasive and charming, and talked freely about other aspects of his life—his interest in art, his wish to write a book about more than a hundred little-known art museums in Europe and to make a virtual online museum of their treasures. He was exuberant, expansive and loquacious when he spoke of all this, and I wondered whether there was an impulsive, "frontal-lobe" flavor to his thinking, as might happen with an incipient frontotemporal dementia. And yet I could not be sure without knowing the patient for a longer time, in depth; perhaps, as his wife insisted, this high-energy ebullience was normal for him.

Recent neurobehavioral testing had revealed that he still showed a tendency to perseveration, impulsivity, inattentiveness in scanning, and memory retrieval deficits—a pattern suggestive, though not diagnostic, of mild frontal lobe and hippocampal dysfunction.

My neurological examination of Mr. K. was unremarkable,

apart from a tremor of his left hand. He had now been off all drugs for several weeks, and his mild parkinsonism had almost entirely disappeared. It was evident, however, that he and his wife were haunted by the uncertainty his physicians had expressed, and which they themselves shared. "Hopefully a steroid psychosis is what it was," Mr. K. said, "but there may have been other underlying causes. Maybe the beginnings of Alzheimer's. What concerns me is that there was no definitive diagnosis. Was it just steroids, or something more serious coming around the bend?" If there was indeed a neurological disease, temporarily unmasked or unleashed by the steroids, was it not still hanging over him, waiting to cause a more irrevocable dementia? Both husband and wife used the term "lurking," and wondered if there was anything more to be done to provide reassurance and a clearer diagnosis.

I could not give the definitive answer they wanted—the whole business was a strange one. There was dispute in the neurological literature as to whether such an entity as "steroid-induced dementia" even existed, and, if so, what its prognosis might be—recovery had been reported in some cases, not in others.

Unable to offer Mr. K. a conclusive diagnosis but reassured by his manifest improvement, I advised him to resume all of his normal activities, hoping that his work, which necessitated much traveling and making complex decisions and negotiations, would give him some reassurance, along with a renewed sense of identity and optimism. When I next saw him, six months later, he told me he had indeed been working very hard: "My illness cost my business dearly. I'm trying to put it back together again."

I followed up with Mr. K. at intervals, and in May 2006—five years after his strange attack of dementia—he scored at a very superior level on tests of mental function across the board. He had recently returned from Europe and Turkey, he told me, and planned to open a business in Dubai. He gave me a fascinating capsule history of the fur trade, and said he intended to move ahead with his online museum.

"Absolutely no carryover from the past," he said. "Almost as if it never happened."

DEMENTIA IS OFTEN SEEN as irreversible—and indeed, in the context of a neurodegenerative disease, it may be so. But there are also dementias so severe as to mimic advanced Alzheimer's disease that may nonetheless be reversible. These are not uncommon in the aging, where inadequate diet and vitamin $B_{12}$ deficiency can lead to neural decline. And to the many possible causes of such reversible dementias—metabolic and toxic disturbances, nutritional imbalances, even psychological stress—excessive use of steroids has to be added. The danger sign, perhaps, is the feeling of extreme well-being they may produce, the euphoria that Mr. K. was so quick to recognize, yet so powerless to resist.

# Tea and Toast

Theresa was in her mid-nineties when admitted to Beth Abraham in 1968. She had slid into a gradually advancing dementia since the age of ninety, though with help from a niece and a visiting nurse, she continued living alone and maintaining a semi-independent existence. But her diet was poor—she lived, her niece told me, on "tea and toast." And now she was becoming confused and incontinent, and needed the care of a nursing home.

She had not apparently had any strokes or seizures, and the default diagnosis was therefore one of "senility," or "SDAT" (senile dementia of Alzheimer type, as we termed it then), a progressive and incurable condition. Outside this there were no abnormalities noted on her general or neurological examination, and routine blood tests showed levels within normal limits. But I was suspicious when I heard of her tea-and-toast diet, and ordered a test that was then somewhat unusual—an assessment of the vitamin B12 level in her serum. The normal range of B12 is between 250 and 1,000 units, but Theresa's blood level, it turned out, was only 45.

This condition, pernicious anemia, is sometimes due to an autoimmune disorder, but more commonly it results from a vegetarian diet. Injections of liver extract were once the standard treatment for such anemia, as it had been observed since the 1920s that eating animal foods, and especially liver, could prevent, halt, or reverse what had been seen as a deficiency disease—though the special factor that made meat, and especially liver, so efficacious was unknown. (George Bernard Shaw, a strict vegetarian, had liver extract injections monthly and, with their help, was able to live to ninety-four, active and creative to the last.)

Repeated attempts to extract the anti-anemia principle in liver were finally successful in 1948—and that same year, when I was fifteen, we had a school visit to the very laboratory where it had been extracted and concentrated from liver (almost as radium had been extracted from pitchblende). We were told that this principle was vitamin B12, or cyanocobalamin, a complex organic compound with a cobalt atom at its center; it had the same beautiful rose-red color characteristic of simple inorganic cobalt salts. This discovery made it possible to test levels of B12 in a patient's blood and to treat the patient, if necessary, with "the red vitamin."[1]

Kinnier Wilson, a neurologist of encyclopedic knowledge, had observed early in the twentieth century that pernicious anemia might, in fact, cause *only* a dementia or a psychosis, without any accompanying anemia, or any neuropathy or spinal cord degeneration—and that such dementia or psychosis might be

---

1. It was only in the 1970s that it became possible, in a great feat of synthetic chemistry, to *synthesize* B12.

largely reversible by liver injections, in contrast to the irreversible structural changes that may occur in the spinal cord when the cause is an autoimmune disorder.[2]

Could this be the case with my old lady? Would her dementia prove reversible if we gave her vitamin B12? To our delight and amazement (for we had thought she might have Alzheimer's disease *in addition to* B12 deficiency), she started to get better with weekly injections of B12. She regained her fluency and memory; she started to go to the hospital library daily, first to look at newspapers and magazines, then to take out books, novels and biographies, the first real reading she had done in nearly five years. She also started back on crossword puzzles, to which she had been addicted. After six months on vitamin B12 injections, she was fully restored and capable of taking charge of her own life and affairs. At this point, she wanted to discharge herself and go back to living at home.

We agreed, though we cautioned her to maintain a full diet, have periodic monitoring of her B12 levels, and take injections as long as they might be needed.

Two years after her discharge from Beth Abraham, Theresa, at ninety-seven, was doing well but still in need of B12 injections. This is the case with many elderly people, whatever their diets, for they often tend to have low gastric acid. (This can be made worse by the medications, such as the proton pump inhibitors,

---

2. Sándor Ferenczi, the great psychoanalyst, started developing some very unusual ideas in the early 1930s—that analysts, for example, should lie down on the analytic couch *beside* their patients. These ideas, albeit a little heretical, were at first seen as expressions of his remarkable originality of mind, but as they grew wilder it became evident that Ferenczi had an organic psychosis, which turned out to be associated with pernicious anemia.

often given for acid reflux, for these can completely prevent the secretion of gastric acid.)

While Theresa was the first, I have now seen similar confusions and dementias due to vitamin B12 deficiency in a number of old people, and it is not always reversible. But Theresa was lucky. "The red vitamin," she said, "it saved my life."

# Telling

Even before my own medical education, I learned an essential truth about doctoring from my parents, both physicians: that being a physician involves much more than handing out diagnoses and treatment; it involves one in some of the most intimate decisions of a patient's life. This requires a considerable amount of human delicacy and judgment, no less than medical judgment and knowledge. If there is a serious, perhaps life-threatening or life-altering condition, what should one tell the patient, and when? How should one tell the patient? *Should* one tell the patient? Every situation is complex, but for the most part, patients want to know the truth, however dire it is. But they want to hear it delivered with tact, with indications if not of hope, then at least a sense of how such life as they have left can be lived in the most dignified, fulfilling way.

Such telling assumes a whole other order of complexity when a patient has a dementia, for here one is intimating not only a death sentence, but one of mental decline, confusion, and, finally, to some degree, loss of self.

. . .

THIS BECAME COMPLEX, tragic, with Dr. M. He had been the medical director of a hospital where I worked, and he had retired at the mandatory age of seventy. But ten years later, in 1982, he came back—this time as a patient with moderately advanced Alzheimer's. He had started to have major difficulties with recent memory, and his wife described him as often confused and disoriented—and sometimes agitated and abusive. She and his doctors had hoped that admitting him to the hospital where he once worked, with surroundings and people he might find familiar, would have a calming and organizing effect on him. I myself and some of the nurses who had worked for Dr. M. were aghast when we heard of this—first that my former chief was now demented, and then that he was to be institutionalized in, of all places, the hospital he had once ruled over as director. This, I thought, would be horribly humiliating, almost an exercise in sadism.

A year after his admission, I summarized his state in a note for his chart:

> I have the melancholic task of seeing my former friend and colleague, now fallen upon such evil days. He was admitted here just a year ago, with the diagnosis . . . of Alzheimer's disease and multi-infarct dementia. . . .
>
> The first weeks and months here were exceedingly difficult. Dr. M. showed incessant "drive" and agitation, and was put on phenothiazines and Haldol to calm him. The effect of these, even in very small doses, was to cause severe lethargy and parkinsonism—he lost weight, he fell constantly, he became cachectic, looked terminal. With the cessation of such drugs, he has regained his physical health

and energy—walks and talks freely, but requires constant attendance (for he would wander off, and is erratic and unpredictable in the extreme). There is striking fluctuation in his mood and mental state—he shows "lucid moments" (or minutes), and returns to his formal, genial personality, but for most of the time is lost in severe disorientation and agitation. Undoubtedly the relation with a devoted attendant is good—and the best we can do. But, unhappily, he is driven and distraught for [much of the time].

It is difficult to know how much he "realizes," and this fluctuates profoundly, almost from second to second.

He enjoys coming to the clinic and yarning of "the old days" with [the nurses]. He seems most at home here, doing this . . . and at such times may be amazingly coherent, able to write (even write prescriptions!).

At such moments, when Dr. M. stepped into his prior role as a hospital director, the transformation was amazingly complete, even if brief. It happened so quickly that none of us quite knew how to react, how to handle this unprecedented situation. But these were rare interludes, I noted, in his frenzied, driven life. In his chart, I wrote:

He is always "on the go," and for much of the time seems to imagine he is still a doctor here; will speak to other patients not as a fellow patient but as a doctor would, and will look through their charts unless stopped.

On one occasion, he saw his own chart, said "Charles M.—that's me," opened it, saw "Alzheimer's disease," and said, "God help me!" and wept.

Sometimes he calls out, "I want to die. . . . Let me die."

Sometimes he fails to recognize Dr. Schwartz, sometimes he fondly calls him "Walter." And I had a very similar experience this morning: when he was brought into [my office], he was very agitated and driven, would not sit down, let me talk with him [or] examine him. A few minutes later, by chance, when I passed him in the corridor, he instantly recognized me (having forgotten, I think, that he had seen me a few minutes before), called me by name, said, "He's the best!" and asked me to help him.

MR. Q. WAS ANOTHER PATIENT, less demented than Dr. M., who resided in a nursing home run by the Little Sisters of the Poor where I often worked. He had been employed for many years as the janitor at a boarding school and now found himself in a somewhat similar place: an institutional building with institutional furniture and a great many people coming and going, especially in the daytime, some in authority, and dressed accordingly, others under their guidance; there was also a strict curriculum, with fixed mealtimes and fixed times for getting up and going to bed. So perhaps it was not entirely unexpected that Mr. Q. should imagine that he was still a janitor, still at a school (albeit a school that had undergone some puzzling changes). But if the pupils were sometimes bedridden or elderly, and the staff wore the white habits of a religious order, these were mere details—he never bothered with administrative matters.

He had his job: checking the windows and doors to make sure they were securely locked at night, inspecting the laundry and boiler room to make sure all was functioning smoothly. The sisters who ran the home, though perceiving his confusion and delusion, respected and even reinforced the identity of this somewhat demented resident, who, they felt, might fall apart if it were taken away. So they encouraged him in his janitorial role, giving him keys to certain closets and encouraging him to lock up at night before he retired. He wore a bunch of keys jangling at his waist—the insignia of his office, his official identity. He would check the kitchen to make sure all of the gas rings and stoves were turned off and no perishable food had been left unrefrigerated. And though he slowly became more and more demented over the years, he seemed to be organized and held together in a remarkable way by his role, the varied tasks of checking, cleaning, and maintenance that he performed throughout the day. When Mr. Q. died of a sudden heart attack, he did so without perhaps ever realizing that he had been anything but a janitor with a lifetime of loyal work behind him.

Should we have told Mr. Q. that he was no longer a janitor but a declining and demented patient in a nursing home? Should we have taken away his accustomed and well-rehearsed identity and replaced it with a "reality" that, though real to us, would have been meaningless to him? It seemed not only pointless but cruel to do so—and might well have hastened his decline.

# The Aging Brain

Having worked as a neurologist in old-age homes and chronic hospitals for almost fifty years, I have seen thousands of older patients with Alzheimer's disease or other dementias, and what strikes me most is the immense diversity of the clinical presentation, despite the fact that most of these patients suffer from disease processes that are pathologically similar. One sees a kaleidoscopic array of symptoms and dysfunctions, never exactly the same in any two people. The neurological dysfunctions interact with all that is particular and unique in an individual—their preexisting strengths and weaknesses, their intellectual powers, their skills, their life experience, their character, their habitual styles, as well as their particular life situations.

Alzheimer's disease may first present as a full-blown syndrome, but more often it starts as isolated symptoms so focal that one may initially suspect a small stroke or tumor; it is only later that the generalized nature of the disease becomes evident (hence the frequent failure to diagnose Alzheimer's at the start). The early symptoms, whether they appear singly or in clusters, are usually subtle. There may be subtle language or memory problems, such as difficulty recalling proper names; subtle perceptual problems, such as momentary illusions or misperceptions; or subtle intellectual problems, such as difficulties getting jokes or following arguments. But in general it is the most recently evolved functions, the complex associational functions, that are the first to be affected.

In these very early stages, the dysfunctions tend to be fugitive and momentary (as is true of the electroencephalographic changes at this time—sometimes one must look through an hour of EEG recordings to find a second of abnormality). But soon there are grosser disturbances of cognition, memory, behavior, judgment, and disorientation in space and time, all finally coalescing as profound global dementia. As the disease advances, sensory and motor disturbances often appear, along with spasticity and rigidity, myoclonus, sometimes seizures, and sometimes parkinsonism. It may bring distressing personality changes and even violent behavior in some people. Finally, there may be almost no responses above a brainstem reflex level. Every possible cortical disorder (and a good many subcortical ones) may be seen in this devastating disease, even though the paths by which the disease advances are so different in each patient.

Sooner or later patients lose the power to articulate their condition, to communicate in any way, except insofar as tone of voice, touch, or music can briefly reach them. Finally, even this is lost and there is indeed total loss of consciousness, of cortical function, of self—psychic death.[1]

---

1. Caring for someone else, especially if that someone is already quite demented and is inexorably going downhill, can involve backbreaking physical exertion as well as a constant, almost telepathic sensitivity to what is going on in a mind now less and less able to communicate its thoughts, less and less able to *have* clear thoughts. People with dementia may get terrifyingly confused and disoriented. Such a burden can make the caretaker ill with stress. As a physician, I see this all too often—sometimes

· · ·

GIVEN THE MULTIFORMITY of symptoms in dementia, one can see why standardized tests, although useful for screening patients and for delineating populations for genetic studies or drug trials, give little idea of what the disease is actually like, the ways in which the beleaguered patient may adapt and react, and the ways in which such people can, on occasion, be helped—or even help themselves.

One of my patients, very early in the course of her disease, suddenly found that she could no longer tell the time when she looked at her watch. She saw the position of the watch hands clearly, but she could not interpret them; for a split second, they made no sense, and then, equally suddenly, they did. These brief visual agnosias rapidly worsened: the unintelligible periods lengthened to seconds, then minutes, and soon the watch hands were unintelligible all the time. She was acutely and mortifyingly conscious of this deterioration; it gave her a sharp sense of horror, of the Alzheimer's process behind it. But she herself was the one to make a crucial therapeutic suggestion: Why don't I wear a digital watch, she asked, and have digital clocks everywhere? She acted on this, and although her agnosia and other problems continued to increase, she remained able to tell the time and organize her day for another three months.

Another of my patients, who was fond of cooking and whose overall cognitive powers were still very good, found that she could no longer compare the volume of liquids in different containers; an ounce of milk did not look the same if it was poured

---

an elderly husband or wife will sacrifice their own health and die before the incapacitated loved one they are caring for; this is why outside help is crucial.

from a glass into a pan, and ludicrous errors started to occur. The patient herself, a former psychologist, ruefully recognized this as a Piagetian error, a loss of the sense of volumetric constancy that is acquired in early childhood. However, by using graduated vessels and measuring cups instead of trying to guess as she used to, she was able to compensate for the problem and to continue safely in the kitchen.

Such patients may perform badly on formal mental testing and yet be able to describe with clarity, vividness, correctness, and humor precisely how one bakes an artichoke or a cake; they may be able to sing a song, tell a story, act a part, play a violin, or paint a painting with remarkably little impairment. It is as if they have lost certain modes of thought while retaining other modes perfectly.

IT IS SOMETIMES SAID that people with Alzheimer's do not realize that they are impaired, that insight is lost from the start. Although this may sometimes be so (if, for example, there is a frontal-lobe type of onset), it is more common, in my experience, for patients to realize their condition at first. Thomas DeBaggio, a writer and horticulturalist, was even able to publish two insightful memoirs about his own early-onset Alzheimer's before the disease killed him at the age of sixty-nine. But most patients are frightened or mortified by the knowledge of what is befalling them. Some continue to be severely terrified as they lose their intellectual competences and bearings and find themselves in a world increasingly fragmented and chaotic. But the majority, I think, become calmer with time as they perhaps start to lose the sense of what they have lost and find themselves shifted into a simpler, unreflective world. Such patients might seem (although one has to beware of this kind of formulation) to have regressed intellectually, so that they are once again like children, restricted to a narrative mode of thought. Kurt Goldstein, a neurologist and psychiatrist, would say of such patients that they had lost not only their abstract capacities but also their abstract "attitude"—that they were now in a lower, more concrete form of consciousness or being.

For Hughlings Jackson, the great English neurologist, there were never just deficits with neurological damage, but "hyper-physiological" or "positive" symptoms, as he called them, "releases" or exaggerations of normally constrained or inhibited neural functions. He spoke of "dissolution," which was, for him, characterized by regression or reversion to more archaic levels of neural function—the reverse of evolution.[2]

Although Jackson's notion of dissolution in the nervous system as evolution in reverse can hardly be maintained now in so simplistic a fashion, one does see some remarkable behavioral regressions or releases in a diffuse cortical disease like Alzheimer's. I have often seen patients with advanced dementia who show picking, hunting, and brushing—a whole range of primitive grooming behaviors that are not seen in normal human development but are suggestive, perhaps, of a phylogenetic reversion to a prehuman, primate level. In the final stages of dementia, where no organized behaviors of any sort remain, one may see reflexes that are normally only seen in infancy, including grasping reflexes, snout and sucking reflexes, and Moro reflexes.

2. Such dissolution was very clear, Jackson thought, in the processes of dreaming, delirium, and insanity, and his long 1894 paper "The Factors of Insanities" is full of fascinating observations and insights in this regard.

One may see remarkable (and sometimes very poignant) behavioral regressions at a more human level, too. I had one patient—a very demented woman of one hundred who was incoherent, distracted, and agitated much of the time—who, if given a doll, would immediately become focused, sharply attentive, and take the doll to her breast as if to nurse it, rocking it in her arms, cuddling it, crooning to it. As long as she was occupied by this mothering behavior, she was perfectly calm; but the moment she stopped, she became agitated and incoherent again.

THE SENSE THAT everything is lost with a diagnosis of Alzheimer's is all too common among neurologists, as well as among patients and their families. This may give rise to a premature sense of impotence and doom, whereas in fact all sorts of neurological functions (including many that subserve the self) seem remarkably able to resist even widespread neuronal dysfunction.

In the early part of the twentieth century, neurologists started to pay more attention not just to the primary symptoms of neurological disease but also to the compensations and adaptations to these. Kurt Goldstein, studying brain-damaged soldiers during World War I, was moved from his original, deficit-based point of view to a more holistic, organismal one. There were never, he believed, just deficits or releases; there were always reorganizations, and these he saw as strategies (albeit unconscious and almost automatic) by which the brain-damaged organism sought to survive, although perhaps in a more rigid and impoverished way.

Ivy Mackenzie, a Scottish physician working with post-encephalitic patients, described the remote effects—"subversions," compensations, and adaptations—that follow the primary insult. In the study of these, he wrote, we see "an organized chaos," ways in which the organism, the brain, comes to terms with itself, reestablishes itself, at other levels. "The physician," he wrote, "is concerned, unlike the naturalist, with a single organism, the human subject, striving to preserve its identity in adverse circumstances."

This theme, the preservation of identity, is well brought out by Donna Cohen and Carl Eisdorfer in their fine book *The Loss of Self,* which is based on painstaking studies of a number of people with Alzheimer's. The title of their book is perhaps misleading, for it is not loss (at least until very late) but surprising preservations and transformations that we see in Alzheimer's, and this, indeed, is what Cohen and Eisdorfer show.[3]

People with Alzheimer's disease may remain intensely human, very much themselves, and capable of normal emotion and relationships until quite late in their illness. (This preservation of self may, paradoxically, be a source of torment for the patient or their families who see them so painfully eroded in other ways.)

The relative preservation of the personal allows a great range of supportive and therapeutic activities that have in common that they address or evoke the personal. Religious services, theater, music and art, gardening, cooking, or other hobbies can anchor patients despite their disintegrations and temporarily restore a focus, an island of identity. Familiar melodies, poems, or stories may still be recognized and responded to despite advanced disease—a response that may be richly associative and bring

---

3. As Henry James was dying, with pneumonia and a high fever, he became delirious—and it is said, as I wrote in *Hallucinations,* that though the master was raving, his style was "pure James" and, indeed, "late James."

back, for a while, some of the patient's memories and feelings and their former powers and worlds. This can bring at least a temporary "awakening" and fullness of life to patients who may otherwise be dismissed or ignored, left in states of bewilderment and vacancy, prone at any moment to losing their bearings or to catastrophic reactions (as Goldstein called them) of unimaginable confusion and panic.

The neural embodiment of self, it seems, is extremely robust. Every perception, every action, every thought, every utterance seems to bear the mark of the individual's experience, of his value system, of all that is peculiar to him. In Gerald Edelman's theory of neuronal group selection (as in Esther Thelen's work on the development of cognition and action in children), we find a rich account of how neuronal connectivity may be determined by, literally shaped by, the individual's experience, thoughts, and actions no less than by all that is hardwired and biologically given. If individual experience and experiential selection so determine the developing brain, we should not perhaps be surprised that individuality, self, is preserved for so long even in the face of diffuse neuronal damage.

AGING, OF COURSE, does not necessarily entail neurological illness. Working in old-age homes, where people are admitted with a variety of problems (heart ailments, arthritis, blindness, or sometimes just loneliness and a desire to live in a community), I see numbers of old people who are, so far as I can judge, intellectually and neurologically wholly intact. Indeed, several of my patients are bright and intellectually active centenarians who have retained all their zest for life, all their interests and faculties, into their eleventh decade. One woman, admitted at the age of 109 with diminishing vision, discharged herself once her cataracts had been taken care of and returned home to an independent life. ("Why should I stay here with all these old people?" she asked.) Even in a chronic hospital, there is a sizable proportion of people who can live out a century or more without significant intellectual decline, and this proportion must be considerably greater in the population at large.

So it is not just the absence of disease or preservation of function that we should be concerned with, but the potential for a continuing development throughout life. Cerebral function is not like cardiac or renal function, which proceeds autonomously, almost mechanically, in a fairly uniform way throughout life. The brain/mind, in contrast, is anything but automatic, for it is always seeking, at every level from the perceptual to the philosophical, to categorize and recategorize the world, to comprehend and give meaning to its own experience. It is the nature of living a real life that experience is not uniform, but ever changing and ever challenging and requiring more and more comprehensive integration. It is not enough for the brain/mind simply to tick over, maintaining uniform function (like the heart); it must adventure and advance throughout life. The very concept of health or wellness requires a special definition in relation to the brain.

A distinction must be made in the aging patient between longevity and vitality. A constitutional sturdiness and good luck may make for a long and healthy life. I think here of five siblings I know, all in their nineties or early hundreds, all looking far younger than their age, and all having the physiques, the sexual drives, the behaviors of much younger people. And yet, human beings may be physically and neurologically healthy but psychically burned out at a relatively early age. If the brain is to stay healthy, it must remain active, wondering, playing, exploring, and experimenting right to the end. Such activities or dispositions may not show up on functional brain imaging or, for that matter, on neuropsychological tests, but they are of the essence in defining the health of the brain and in allowing its development throughout life. This is clear in Edelman's neurobiological model, where the brain/mind is conceived as incessantly active, categorizing and recategorizing its activities throughout life, constructing interpretations and meanings at ever higher levels.

Such a neurobiological model accords well with what Erik and Joan Erikson devoted a lifetime to studying: universal, age-related stages that seem to appear in all cultures. As the Eriksons themselves advanced through their nineties, they added a further stage to the eight stages they originally described. This last stage is well recognized and respected in many cultures (although sometimes forgotten in our own). This is the stage appropriate to old age; and the solution or strategy to be achieved at this stage is what the Eriksons call wisdom or integrity.

The achievement of this stage involves the integration of vast amounts of information, the synthesis of a long lifetime's experience, coupled with the lengthening and enlargement of the individual's perspectives and a sort of detachment or calm. Such a process is entirely individual. It cannot be prescribed or taught; nor is it directly dependent on education or intelligence or specific talents. "We cannot be taught wisdom," as Proust remarks, "we have to discover it for ourselves by a journey which no one can undertake for us, an effort which no one can spare us."

Are such stages purely existential or cultural—the behaviors, the perspectives appropriate to various ages and stages—or do they also have some specific neural basis? We know that learning is possible throughout life, even in the presence of cerebral aging or disease, and we can be sure that other processes, at a much deeper level, are continuing, too—a culmination of the ever wider and deeper generalizations and integrations that have been occurring in the brain/mind throughout life.

In the nineteenth century, when a powerful mind could still take all of nature for its subject, the great naturalist Alexander von Humboldt, after a lifetime of travel and scientific research, embarked in his mid-seventies on a grand synthetic view of the universe, bringing together everything he had seen and thought into a final work, *Cosmos*. He was well into its fifth volume when he died at eighty-nine. In our own time, when even the largest minds must narrow their gaze, the evolutionary biologist Ernst Mayr recently gave us, in his ninety-third year, *This Is Biology*, a marvelous book on the rise and scope of biology that combines the spaciousness that a lifetime of thought brings with the eager immediacy of the boy who passionately tracked birds eighty years before. Such passion, as Mayr writes, is the key to vitality in old age:

> The most important ingredient [for a biologist] is a fascination with the wonders of living creatures. And this stays with most biologists for their entire life. They never lose the excitement of scientific discovery . . . nor the love of chasing after new ideas, new insights, new organisms.

If we are lucky enough to reach a healthy old age, this sense of wonder can keep us passionate and productive to the end of our lives.

# Kuru

In 1997, I saw a patient in New York, an eighty-seven-year-old woman who had been physically active, intellectually intact, and seemingly in good health until the beginning of that year. In the last days of January, however, she became strangely excited, then agitated—"Something terrible is happening to me," she said. It was difficult for her to sleep: the curtains and the corners of the room seemed peopled by ghostly faces, and what sleep she had was broken by vivid dreams. On the fifth day, periods of confusion and disorientation began to appear. A medical disturbance was suspected—perhaps a urinary infection, a chest infection, some toxic or metabolic disturbance—but her attending physician could find no fever, nor any abnormality in her blood or urine. CT scans of her brain appeared normal. A psychiatric opinion was sought—depression in the elderly can sometimes present itself as confusion—but this notion became increasingly untenable as, within days, the initial confusion deepened.

By the middle of February, myoclonic jerks of the muscles convulsed her limbs, her abdomen, and her face. Her speech lost coherence and intelligibility from day to day, and an increasing spasticity took hold. By the third week of her illness, she no longer recognized her own children.

Towards the end of the month, she started to alternate between states of stuporlike sleep and a restless, twitching delirium, in which a light touch might bring on violent jerks of her whole body. She died on March 11, emaciated, rigid, in a coma, less than six weeks after the onset of her first symptoms. We sent a tissue sample from her brain to the pathologist, for the overwhelming possibility was that she had Creutzfeldt-Jakob disease. The pathologist, indeed, was visibly uneasy; no pathologist is quite comfortable dealing with the tissues of such patients.

Neurologists see incorrigible diseases all the time, and yet this case threw me to an unusual degree, with its devastating clinical course, the almost visible day-to-day destruction of the brain, the racking myoclonic spasms of the body, and our manifest inability to do anything for the patient.

Creutzfeldt-Jakob disease is rare—its incidence is about one in a million per year—and I had seen it only once before, in 1964, when I was a neurology resident. The unfortunate man afflicted with it then was presented to us as a case of a highly unusual degenerative brain disease. There was discussion of its typical features: a rapidly advancing dementia; sudden, lightning myoclonic jerks of the muscles; and a peculiar-looking "periodic" electroencephalogram. These, we were told, constituted the diagnostic triad for CJD. Only about twenty cases had been reported since the original account by Creutzfeldt and Jakob in 1920, and we were excited at encountering such a neurological rarity. At that time, neurology was still largely descriptive, almost ornithological, and CJD was seen as a rara avis, along with Hallervorden-Spatz disease, Unverricht-Lundborg syndrome, and other such exotic, eponymous rarities.

We had no idea in 1964 of the truly singular nature of CJD, its affinities with other human and animal diseases, or that it would turn out to be the archetype, the epitome, of a whole new order of disease. We never for a moment thought of it as infectious; indeed, we took blood and spinal fluid from this patient nonchalantly, as from any other, not dreaming that an inadvertent needle stick, the accidental implanting of a grain of tissue, might cause us to share his fate. It was only in 1968 that CJD was shown to be a transmissible disease.

IN 1957, CARLETON GAJDUSEK, a brilliant young American physician and ethologist who had already done notable work exploring disease "isolates" elsewhere in the world, went to New Guinea to investigate a mysterious neurological disease decimating the villages of the Fore people. It seemed to affect women and children almost exclusively, and it had apparently never occurred before the present century. The Fore called it "kuru" and ascribed it to sorcery. The clinical course of kuru was one of rapid and remorseless neurological deterioration—falling, staggering, paralyses, and involuntary laughter—ending fatally within a few months. The brains of those who died showed devastating changes, some areas having been reduced to a virtual sponge, riddled with holes. The cause of this disease was exceedingly puzzling—genetic factors, toxic factors, disease agents of the usual sort were all considered and found irrelevant. It required original work, much of it in the difficult field conditions of western New Guinea, for Gajdusek to trace the disease to the transmission of a new kind of agent, one that could rest for years in the tissues of those affected without causing any symptoms and then, after this huge latent period, begin a rapidly fatal process. He used the term "slow virus" for this singular agent, and he demonstrated that it was the practice of cannibalism as part of funeral rites (specifically, the eating of infected brain tissue) that caused its spread among the Fore. He went on to show that the agent could cause a similar disease when it was given to chimps and monkeys. For this work he received a Nobel Prize in 1976.

Richard Rhodes, in his 1997 book *Deadly Feasts,* has told the kuru story with psychological insight and dramatic force, virtually reliving the early days of this investigation—a time of fear, bewilderment, great ambition, and intellectual discovery.

From its prelude in New Guinea, Rhodes's chronicle opens out, steadily, to wider and wider prospects, showing how connections with other human diseases and with various animal diseases were laboriously pieced together. Not the least of his book's many excellences is Rhodes's depiction of the great role that chance and luck and unexpected encounters play in the all too human business of science. A most crucial serendipity came in 1959, when an English veterinarian, William Hadlow, saw an exhibit of photographs—Gajdusek's "kuru show"—then on display in the Wellcome Historical Medical Museum in London. Hadlow was instantly struck by the similarities of the clinical and pathological picture of kuru to that of a fatal sheep disease, scrapie, which had affected isolated flocks of sheep in England and elsewhere since the early eighteenth century. (It had been endemic in Central Europe before that, and spread to the United

States in 1947.) And scrapie, as Hadlow pointed out in a letter to *The Lancet,* was known to be transmissible. Gajdusek had briefly thought about an infectious basis for kuru, but had then discarded the notion; now he was forced into reconsidering it—indeed, into seeing that kuru *must* be infectious, and that any similar human diseases would almost certainly be infectious, too. Demonstrating this experimentally took years of patient and difficult work that involved injecting chimps with kuru- and CJD-infected tissue—work made more difficult by the long incubation periods of these diseases.

All of these diseases—kuru, CJD, scrapie, and various rarer ones like fatal familial insomnia and Gerstmann-Sträussler-Scheinker syndrome—are relentlessly progressive and rapidly fatal. All produce devastating spongy changes and cavitations in the brain—hence they are referred to collectively as the transmissible spongiform encephalopathies, or TSEs. The disease agents in each of these are very difficult to isolate, smaller than viruses, and, ominously, capable of surviving the most drastic conditions, including extreme heat and pressure as well as chemicals like formaldehyde and all the usual sterilization procedures.

Bacteria are autonomous and multiply by themselves; viruses use their genetic material to subvert the host's cells to replicate—but the TSE disease agents show no evidence of containing any RNA or DNA at all. How, then, could they be characterized, and how could they cause disease? Gajdusek named these agents "infectious amyloids." (They are now known as "prions," the name given to them by Stanley Prusiner, who was awarded a Nobel Prize for his work in identifying this new class of pathogen.) But if prions could not replicate like viruses, how did they multiply and spread? One had to envisage a wholly new form of disease process—one akin not to biological replication but to chemical crystallization, whereby the tiny prions, which are actually deviant, pleated forms of a normally present brain protein, act as "pattern-setting nucleants," or centers of recrystallization, causing rapidly spreading transformations of the surrounding crystalline proteins. Such nucleation is seen in the patterning of ice or snowflakes, and an apocalyptic form of this was imagined years ago by Kurt Vonnegut in *Cat's Cradle*, where the world is ended by a sliver of a substance that transforms all water into unmeltable "ice-nine."[1]

Prions infect us not by lodging in us as invaders but by evoking a disturbance in our own brain proteins. It is for this reason that there is no inflammation or immune reaction to the prions, for our own proteins, normal or abnormal, are not perceived by the immune system as foreign. It is the helplessness of the organism in the face of its own subversion, allied to the near indestructibility of prions, that makes the TSEs potentially the deadliest diseases on earth. While the TSEs may be extremely rare in nature, arising only from a very occasional, stochastic transformation of brain protein (this seems to account for the remarkably constant one-in-a-million incidence of sporadic CJD throughout the world each year), cultural practices—eating brains or feeding offal or animal remains to cattle—may radically alter the picture and cause a galloping transmission of these diseases.

---

1. Prions were seen first as "slow" viruses, then as "unconventional" ones, but if we categorize them as "viruses" or "alive," we must radically redefine what we mean by either term, for they seem to belong, in many ways, to a purely crystalline world. (Gajdusek, indeed, entitled one of his early papers "Fantasy of a 'Virus' from the Inorganic World.")

. . .

IN THE EARLY DAYS it was often thought that kuru was no more than, as Rhodes puts it, "a tragic curiosity" confined to a few Stone Age cannibals at the other end of the earth. But Gajdusek insisted from the first on its much wider potential relevance. It was Gajdusek and his colleagues at the National Institutes of Health who showed, in 1968, that CJD was a transmissible spongiform encephalopathy, like kuru, and who warned that this disease might be transmitted accidentally by surgical or dental procedures. Exactly this happened in the early 1970s, following corneal transplantation in one patient and neurosurgery (with autoclaved but still infected instruments) in others.

A greatly heightened incidence of disease was seen in the 1990s in patients who had received human growth hormone (extracted from cadaveric pituitary glands) as children: of some 11,600 patients who were given it, at least 86 developed CJD. Fortunately, synthetic growth hormone became available in the mid-1980s, forestalling further disasters.

Around that time, too, a new disease appeared among some cattle in Britain, causing them to act strangely, stagger, and rapidly die. It was dubbed, popularly, "mad cow disease"; scientists referred to it as bovine spongiform encephalopathy, or BSE. Cattle, of course, are normally vegetarians, but increasingly they have been fed on a high-protein meat-and-bonemeal mix, a slaughterhouse by-product sometimes containing, among other things, the offal of diseased cattle and sheep—perhaps including brain tissue from scrapie-infested sheep. Whether the cannibalistic eating of cattle brains amplified a previously rare and sporadic disease (like the cannibalism among the Fore did) or whether scrapie prions from sheep crossed the species barrier to infect cattle is unclear. But the use of meat-and-bonemeal feed soon brought about a disaster.

More than a dozen young people died of a CJD variant in Britain in the late 1990s, and it seems likely that they acquired their disease from eating infected meat products. The clinical picture in these cases—early behavioral changes and incoordination—was more reminiscent of kuru than of "classical" CJD (and this was true of the pathological changes, too).

But the incubation period, as the Fore people showed, can be many decades, and in the United States and elsewhere there are large reservoirs of TSE in sheep and mink, as well as in some wild deer and elk, and continuing use of meat-and-bonemeal feed for pigs, chickens, and cattle. And it may be, as Gajdusek theorized, that no source of food can be considered safe from infection by prionlike agents. Meat-and-bonemeal waste and animal by-products are even used sometimes to fertilize organic vegetable crops, and animal fat and gelatin are widely used in food, cosmetics, and pharmaceuticals.

Such practices have now been banned in several countries.

# A Summer of Madness

O n July 5, 1996," Michael Greenberg starts, "my daughter
was struck mad." No time is wasted on preliminaries, and
his memoir *Hurry Down Sunshine* moves swiftly, almost tor-
rentially, from this opening sentence, in tandem with the events
it tells of. The onset of mania is sudden and explosive: Sally,
Greenberg's fifteen-year-old daughter, has been in a heightened
state for some weeks, listening to Glenn Gould's *Goldberg Variations* on her Walkman, poring over a volume of Shakespeare's
sonnets till the early hours. Greenberg writes:

> Flipping open the book at random I find a blinding crisscross
> of arrows, definitions, circled words. Sonnet 13 looks like a
> page from the Talmud, the margins crowded with so much
> commentary the original text is little more than a speck at
> the center.

Sally has also been writing singular, Sylvia Plath–like poems.
Her father surreptitiously glances at these and finds them strange,
but it does not occur to him that her mood or activity is in any

way pathological. She has had learning difficulties from an early age, but she is now triumphing over these, finding her intellectual powers for the first time. Such exaltation is normal in a highly gifted fifteen-year-old. Or so it seems.

But on that hot July day, she breaks—haranguing strangers in the street, demanding their attention, shaking them, and then suddenly running full tilt into a stream of traffic, convinced that she can bring it to a halt by sheer willpower. (With quick reflexes, a friend yanks her out of the way just in time.)

In an unpublished draft of his *Life Studies,* Robert Lowell described something very similar in an attack of "pathological enthusiasm":

> The night before I was locked up I ran about the streets of Bloomington Indiana. . . . I believed I could stop cars and paralyze their forces by merely standing in the middle of the highway with my arms outspread.

Such sudden, dangerous exaltations and actions are not uncommon at the start of a manic attack.

Lowell had a vision of evil in the world, and of himself, in his "enthusiasm," as the Holy Ghost. Sally had, in some ways, an analogous vision of moral collapse, seeing all around herself the loss or suppression of God-given "genius," and of her own mission to help everyone reclaim that lost birthright. That it was such a vision which led to her passionate confrontation with strangers, her bizarre behavior imbued with a sense of her own special powers, her parents learn when they quiz her the next day. Greenberg writes:

She has had a vision. It came to her a few days ago, in the Bleecker Street playground, while she was watching two little girls play on the wooden footbridge near the slide. In a surge of insight she saw their genius, their limitless native little-girl genius, and simultaneously realized that we are all geniuses, that the very idea the word stands for has been distorted. Genius is not the fluke they want us to believe it is, no, it's as basic to who we are as our sense of love, of God. Genius is childhood. The Creator gives it to us with life, and society drums it out of us before we have the chance to follow the impulses of our naturally creative souls. . . .

Sally related her vision to the little girls in the playground. Apparently they understood her perfectly. Then she walked out onto Bleecker Street and discovered her life had changed. The flowers in front of the Korean deli in their green plastic vases, the magazine covers in the news shop window, the buildings, cars—all took on a sharpness beyond anything she had imagined. The sharpness, she said, "of present time." A wavelet of energy swelled through the center of her being. She could see the hidden life in things, their detailed brilliance, the funneled genius that went into making them what they are.

Sharpest of all was the misery on the faces of the people she passed. She tried to explain her vision to them but they just kept rushing by. Then it hit her: they already know about their genius, it isn't a secret, but much worse: genius has been suppressed in them, as it had been suppressed in her. And the enormous effort required to keep it from percolating to the surface and reasserting its glorious hold on our lives is the cause of all human suffering. Suffering that Sally, with this epiphany, has been chosen among all people to cure.

As startling as Sally's passionate new beliefs are, her father and stepmother are even more struck by her manner of speaking:

> Pat and I are dumbstruck, less by what she is saying than how she is saying it. No sooner does one thought come galloping out of her mouth than another overtakes it, producing a pile-up of words without sequence, each sentence canceling out the previous one before it's had a chance to emerge. Our pulses racing, we strain to absorb the sheer volume of energy pouring from her tiny body. She jabs at the air, thrusts out her chin . . . her drive to communicate is so powerful it's tormenting her. Each individual word is like a toxin she must expel from her body.
>
> The longer she speaks, the more incoherent she becomes, and the more incoherent she becomes, the more urgent is her need to make us understand her! I feel helpless watching her. And yet I am galvanized by her sheer aliveness.

One may call it mania, madness, or psychosis—a chemical imbalance in the brain—but it presents itself as energy of a primordial sort. Greenberg likens it to "being in the presence of a rare force of nature, such as a great blizzard or flood: destructive, but in its way astounding too." Such unbridled energy can resemble that of creativity or inspiration or genius—this, indeed, is what Sally feels is rushing through her: not an illness, but the apotheosis of health, the release of a deep, previously suppressed self.

These are the paradoxes that surround what Hughlings Jackson, the nineteenth-century neurologist, called "super-positive" states: they betoken disorder, imbalance in the nervous system, but their energy, their euphoria, makes them feel like supreme health. Some patients may achieve a startled insight into this, as did one patient of mine, a very old lady with neurosyphilis. Becoming more and more vivacious in her early nineties, she said to herself, "You're feeling too well, you must be ill." George Eliot, similarly, spoke of herself as feeling "dangerously well" before the onset of her migraine attacks.

Mania is a biological condition that feels like a psychological one—a state of mind. In this way it resembles the effects of various intoxications. I saw this very dramatically with some of my *Awakenings* patients when they began taking L-dopa, a drug that is converted in the brain to the neurotransmitter dopamine. Leonard L., in particular, became quite manic on this: "With L-dopa in my blood," he wrote at the time, "there's nothing in the world I can't do if I want." He called dopamine "resurrectamine" and started to see himself as a messiah—he felt that the world was polluted with sin and that he had been called upon to save it. And in nineteen nonstop, almost sleepless days and nights, he typed an entire autobiography of fifty thousand words. "Is it the medicine I am taking," wrote another patient, "or just my new state of mind?"

If there is uncertainty in a patient's mind about what is "physical" and what is "mental," there may be a still deeper uncertainty as to what is self or not-self—as with my patient Frances D., who, as she grew more excited on L-dopa, was taken over by strange passions and images that she could not dismiss as entirely alien to her "real self." Did they, she wondered, come from very deep but previously suppressed parts of herself? But these patients, unlike Sally, knew that they were on a drug, and could see, all around them, similar effects taking hold on the others.

For Sally there was no precedent, no guide. Her parents were as bewildered as she was—more so, because they did not have her mad assurance. Was it, they wondered, something she had been taking—had she dropped acid, or something worse? And if not, was it something they had bequeathed her in their genes or something awful they had "done" at a critical stage in her development? Was it something she had always had in her, even though it triggered so suddenly?

These were the questions my own parents asked themselves when, in 1943, my fifteen-year-old brother Michael became acutely psychotic. My brother saw "messages" everywhere, felt his thoughts were being read or broadcast, had explosions of strange giggling, and believed he had been translocated to another "realm." Hallucinatory drugs were rare in the 1940s, so my parents, who were both doctors, wondered whether Michael might have some psychosis-producing illness—perhaps a thyroid condition or a brain tumor. It ultimately became clear, though, that my brother suffered from a schizophrenic psychosis. In Sally's case, blood tests and physical exams ruled out any problems with thyroid levels, intoxicants, or tumors. Her psychosis, though acute and dangerous (all psychoses are potentially dangerous, at least to the patient), was "merely" manic.

One can become manic (or depressed) without becoming psychotic—that is, having delusions or hallucinations, losing sight of reality. Sally, though, did go over the top, and on that hot July day, something happened, something snapped. All of a sudden, she was a different person—she looked different, sounded different. "Suddenly every point of connection between us had vanished," her father writes. She calls him "Father" (he was "Dad" before) and speaks in a "pressured, phony voice, as if delivering stage lines she has learned"; "her normally warm chestnut eyes are shell-like and dark, as if they've been brushed over with lacquer."

Greenberg tries to talk with her of ordinary matters, asking her if she is hungry or wants to lie down:

> Each time, however, her otherness is reaffirmed. It is as if the real Sally has been kidnapped, and here in her place is a demon, like Solomon's, who has appropriated her body. The ancient superstition of possession! How else to come to grips with this grotesque transformation? . . . In the most profound sense Sally and I are strangers: we have no common language.

The special qualities of mania have been recognized and distinguished from other forms of madness since the great physicians of antiquity wrote on the subject. The Greek physician Aretaeus, in the second century, gave a clear description of how excited and depressed states might alternate in an individual, but the distinction between different forms of madness was not formalized until the rise of psychiatry in nineteenth-century France. It was then that "circular insanity" (*folie circulaire* or *folie à double forme*)—what Emil Kraepelin later called manic-depressive insanity and what we would now call bipolar disorder—was distinguished from the much graver disorder of "dementia praecox," or schizophrenia. But medical accounts, accounts from the outside, can never do justice to what is actually experienced in the course of such psychoses; there is no substitute here for firsthand accounts.

There have been several such personal narratives over the years, and one of the best, to my mind, is *Wisdom, Madness and Folly: The Philosophy of a Lunatic,* by John Custance, published in 1952. He writes:

> The mental disease to which I am subject is . . . known as manic depression, or, more accurately, as Manic-depressive Psychosis. . . . The manic state is one of elation, of pleasurable excitement sometimes attaining to an extreme pitch of ecstasy; the depressive state is its precise opposite, one of misery, dejection, and at times of appalling horror.

Custance had his first manic attack at the age of thirty-five and would continue to have periodic episodes of mania or depression for the next twenty years:

> When the nervous system is thoroughly deranged, the two contrasting states of mind can be almost infinitely intensified. It sometimes seems to me as though my condition had been specially devised by Providence to illustrate the Christian concepts of Heaven and Hell. Certainly it has shown me that within my own soul there are possibilities of an inner peace and happiness beyond description, as well as of inconceivable depths of terror and despair.
>
> Normal life and consciousness of "reality" appear to me rather like motion along a narrow strip of table-land at the top of a Great Divide separating two distinct universes from

each other. On the one hand the slope is green and fertile, leading to a lovely landscape where love, joy and the infinite beauties of nature and of dreams await the traveller; on the other a barren, rocky declivity, where lurk endless horrors of distorted imagination, descends to the bottomless pit.

In the condition of manic-depression, this table-land is so narrow that it is exceedingly difficult to keep on it. One begins to slip; the world about one changes imperceptibly. For a time it is possible to keep some sort of grip on reality. But once one is really over the edge, once the grip of reality is lost, the forces of the Unconscious take charge, and then begins what appears to be an unending voyage into the universe of bliss or the universe of horror as the case may be, a voyage over which one has oneself no control whatever.

In our own time, Kay Redfield Jamison, a brilliant and courageous psychologist who has manic-depressive illness herself, has written both the definitive medical monograph on this subject (*Manic-Depressive Illness*, with Frederick K. Goodwin) and a personal memoir, *An Unquiet Mind*. In the latter, she writes:

I was a senior in high school when I had my first attack of manic-depressive illness; once the siege began, I lost my mind rather rapidly. At first, everything seemed so easy. I raced about like a crazed weasel, bubbling with plans and enthusiasms, immersed in sports, and staying up all night, night after night, out with friends, reading everything that wasn't nailed down, filling manuscript books with poems and fragments of plays, and making expansive, completely unrealistic, plans for my future. The world was filled with pleasure and promise; I felt great. Not just great, I felt *really* great. I felt I could do anything, that no task was too diffi-cult. My mind seemed clear, fabulously focused, and able to make intuitive mathematical leaps that had up to that point entirely eluded me. Indeed, they elude me still.

At that time, however, not only did everything make per-fect sense, but it all began to fit into a marvelous kind of cosmic relatedness. My sense of enchantment with the laws of the natural world caused me to fizz over, and I found myself buttonholing my friends to tell them how beautiful it all was. They were less than transfixed by my insights into the webbings and beauties of the universe, although consid-erably impressed by how exhausting it was to be around my enthusiastic ramblings. . . . Slow down, Kay. . . . For God's sake, Kay, slow down.

I did, finally, slow down. In fact, I came to a grinding halt.

Jamison contrasts this experience with her later bouts:

Unlike the very severe manic episodes that came a few years later and escalated wildly and psychotically out of control, this first sustained wave of mild mania was a light, lovely tincture of true mania. . . . It was short-lived and quickly burned itself out: tiresome to my friends, perhaps; exhaust-ing and exhilarating to me, definitely; but not disturbingly over the top.

Both Jamison and Custance describe how mania alters not just thought and feeling, but even their sensory perceptions. Custance carefully itemizes these changes in his memoir. Some-

times the electric lights in the ward have "a bright starlike phe-
nomenon emanat[ing] . . . ultimately forming a maze of iridescent
patterns." Faces seem to "glow with a sort of inner light which
shows up the characteristic lines extremely vividly." Though normally "a hopeless draughtsman," Custance is able to draw quite
well while manic (I was reminded here of my own ability to do
this, many years ago, during a period of amphetamine-induced
hypomania); all of his senses seem intensified:

My fingers are much more sensitive and neat. Although
generally a clumsy person with an execrable handwriting I
can write much more neatly than usual; I can print, draw,
embellish and carry out all sorts of little manual operations,
such as pasting up scrapbooks and the like, which would
normally drive me to distraction. I also note a particular
tingling in my fingertips.

My hearing appears to be more sensitive, and I am able
to take in . . . many different sound-impressions at the same
time. . . . From the cries of gulls outside to the laughter and
chatter of my fellow-patients, I am fully alive to what is
going on and yet find no difficulty in concentrating on my
work.

. . . If I were to be allowed to walk about freely in a
flower garden I should appreciate the scents far more than
usual. . . . Even common grass tastes excellent, while real
delicacies like strawberries or raspberries give ecstatic sensations appropriate to a veritable food of the gods.

At first, Sally's parents struggle to believe (as Sally herself
believes) that her excited state is something positive, something

other than illness. Her mother puts a somewhat New Agey spin on it:

> Sally is having an experience, Michael, I'm sure of it, this isn't a sickness. She's a highly spiritual girl. . . . What's happening right now is a necessary phase in Sally's evolution, her journey toward a higher realm.

And this interpretation finds echoes of a more classical kind in Greenberg himself:

> I wanted to believe this too . . . to believe in her breakthrough, her victory, the delayed efflorescence of her mind. But how does one tell the difference between Plato's "divine madness" and gibberish? between [enthusiasm] and lunacy? between the prophet and the "medically mad"?

(It was similar, Greenberg points out, with James Joyce and his schizophrenic daughter, Lucia. "Her intuitions are amazing," Joyce remarked. "Whatever spark of gift I possess has been transmitted to her and has kindled a fire in her brain." Later, he told Beckett, "She's not a raving lunatic, just a poor child who tried to do too much, to understand too much.")

But it becomes clear within hours that Sally is indeed psychotic and out of control, and her parents take her to a psychiatric hospital. At first, she welcomes this, seeing the nurses, the attendants, and the psychiatrists as specially tuned to understand her insight, her message. The reality is brutally different: she is stupefied with tranquilizers and put in a locked ward.

Greenberg's description of the ward takes on the richness and density of a novel, embracing a Chekhovian cast of characters—the staff, the other patients on the ward. He sees a highly disturbed, obviously psychotic young Hasidic man whose family will not accept that he is ill: "He has achieved *devaykah*," says his brother, "the state of constant communion with God."

There is relatively little attempt to *understand* Sally in the hospital—her mania is treated first of all as a medical condition, a disturbance of brain chemistry, to be dealt with on a neurochemical basis. Medication is crucial, even lifesaving, in acute mania, which untreated can lead to exhaustion and death. Unfortunately, though, Sally does not respond to lithium, which has been invaluable for many patients with manic-depressive illness, and so her physicians have to resort to heavy tranquilizers—which damp down her exuberance and wildness but leave her stupefied and apathetic and parkinsonian for a time. Seeing his teenage daughter in this zombielike state is almost as shocking for her father as her mania has been.

AFTER TWENTY-FOUR DAYS of this, Sally is released—still somewhat delusional and still on strong tranquilizers—to go home, under careful and at first continuous surveillance. Outside the hospital, she establishes a crucial relationship with an exceptional therapist, who is able to approach her as a human being, trying to understand her thoughts and feelings. Dr. Lensing shows a disarming directness: "I bet you feel as if there's a lion inside you" are her first words to Sally.

"How did you know?" Sally is amazed, her suspicion instantly melting away. Lensing goes on to talk of mania, Sally's mania, as if it were a sort of creature, another being, inside her:

Lensing nimbly lowers herself into the waiting area chair next to Sally's and tells her in a tone of woman-to-woman straight talk that mania—and she refers to it as if it is a separate entity, a mutual acquaintance of theirs—mania is a glutton for attention. It craves thrills, action, it wants to keep thriving, it will do anything to live on. "Did you ever have a friend who's so exciting you want to be around her, but she leads you into disaster and in the end you wish you never met? You know the sort of person I mean: the girl who wants to go faster, who always wants more. The girl who serves herself first and screw the rest. . . . I'm just giving an example of what mania is: a greedy, charismatic person who pretends to be your friend."

Lensing tries to get Sally to distinguish her psychosis from her true identity, to stand outside the psychosis and to see the complex, ambiguous relationship between it and her. (Psychosis is "not an identity," she says sharply.) She speaks of this to Sally's father, too—for his understanding is also necessary if Sally is to get better. She emphasizes the seductive power of psychosis:

"Sally . . . doesn't want to be isolated, her impulse is outward, which I can tell you is extremely good news. Her desire is to be understood, and not only by us, she wants to understand herself as well. She's still attached to her mania, of course. She's remembering the intensity of her experience, and she's doing her damnedest to keep that intensity alive. She thinks that if she gives it up, she'll lose the great abilities she believes she's acquired. It's a terrible paradox really: the mind falls in love with psychosis. The evil seduction, I call it."

"Seduction" is the crucial word here (it is also the key word in the title of Edward Podvoll's marvelous book *The Seduction of Madness,* on the nature and treatment of mental illness). Why should psychosis, and mania in particular, be seductive? Freud spoke of all psychoses as narcissistic disorders: one becomes the most important person in the world, chosen for a unique role, whether it is to be a messiah, a redeemer of souls, or (as happens in depressive or paranoid psychoses) to be the focus of universal persecution and accusation or derision and degradation.

But even short of such messianic feelings, mania can fill one with a sense of enormous pleasure, even ecstasy—and the sheer intensity of this may make it difficult to "give up." It is what prompts Custance, despite his knowledge of how dangerous such a course is, to avoid medication and hospitalization in one attack of mania and, instead, embrace it, undertaking a risky and rather James Bond–like adventure in East Berlin. Perhaps a similar intensity of feeling is sought by drug addicts, especially those addicted to stimulants like cocaine or amphetamines; and here, too, a high is likely to be followed by a crash, just as a mania is usually followed by a depression—both, perhaps, due to the exhaustion caused by neurotransmitters like dopamine in the overstimulated reward systems in the brain.

Mania, though, is by no means all pleasure, as Greenberg continually observes. He speaks of Sally's "pitiless ball of fire," her "terrified grandiosity," of how anxious and fragile she is inside the "hollow exuberance" of her mania. When one ascends to the exorbitant heights of mania, one becomes very isolated from ordinary human relationships, human scale—even though this isolation may be covered over by a defensive imperiousness or grandiosity. This is why Lensing sees Sally's returning desire to make genuine contact with others, to understand and be understood, as a propitious sign of her returning to health, her coming back to earth.

Psychosis, as Lensing says, is not an identity, but a temporary aberration or departure from identity. And yet having a chronic or recurring mind-altering condition like manic-depressive illness is bound to influence one's identity, to become part of one's attitudes and ways of thinking. As Jamison writes,

> It is, after all, not just an illness, but something that affects every aspect of my life: my moods, my temperament, my work, and my reactions to almost everything that comes my way.

Nor is it just a piece of biological bad luck. Although Jamison agrees that there is nothing good to be said for depression, she does feel that her manias and hypomanias, when not too out of control, have played a crucial and sometimes positive part in her life. Indeed, in her book *Touched with Fire: Manic-Depressive Illness and the Artistic Temperament,* she has provided much evidence to suggest a possible relationship between mania and creativity, citing the many great artists—Schumann, Coleridge, Byron, and Van Gogh among them—who seem to have lived with manic-depressive illness.

When Sally is hospitalized, her father asks the psychiatric resident about her diagnosis. "Sally's condition," the resident says, "has probably been building for a while, gathering strength until it just overwhelmed her." Greenberg asks what her "condition" is. He is told, "What we call [it] is not what's important right now. Certainly many of the criteria for bipolar 1 are here. But fifteen is relatively early for fulminating mania to present itself."

In the last couple of decades, the term "bipolar disorder" has come into use, in part, Jamison suggests, because it is felt to be less stigmatizing than "manic-depressive illness." "But," she cautions,

> splitting mood disorders into bipolar and unipolar categories presupposes a distinction between depression and manic-depressive illness . . . that is not always clear, nor supported by science. Likewise, it perpetuates the notion that depression exists rather tidily segregated on its own pole, while mania clusters off neatly and discreetly on another. This polarization . . . flies in the face of everything that we know about the cauldronous, fluctuating nature of manic-depressive illness.

Moreover, "bipolarity" is characteristic of many disorders of control—like catatonia and parkinsonism—where patients lose the middle ground of normality and alternate between hyperkinetic and akinetic states. Even in a metabolic disease such as diabetes, there may be dramatic alternations between (for instance) very high blood sugar and very low blood sugar, as the complex homeostatic mechanisms are compromised.

There is another reason why the notion of manic-depressive illness as a bipolar illness, swinging from one pole to the other, can be misleading. This was brought out by Kraepelin more than a century ago, when he wrote of "mixed states," states in which there are elements of both mania and depression, inseparably intertwined. He wrote of "the deep inward relationship of such apparently contradictory states."

We speak of "poles apart," but the poles of mania and depression are so close to each other that one wonders if depression may be a form of mania, or vice versa. (Such a dynamic notion of mania and depression—their "clinical unity," as Kraepelin put it—is underlined by the fact that lithium, for those patients in whom it works, works equally well on *both* states.) This paradoxical situation is described by Greenberg with often astonishing oxymorons, as when he speaks of the "abysmal elation" Sally sometimes feels "in the throes of [her] dystopic mania."

SALLY'S FINAL RETURN from the mad heights of her mania is almost as sudden as her taking off into it seven weeks earlier, as Greenberg recounts:

> Sally and I are standing in the kitchen. I have spent the day at home with her, working on my script for Jean-Paul.
>
> "Would you like a cup of tea?" I ask. "That would be nice. Yes. Thank you." "With milk?" "Please. And honey."
>
> "Two spoonfuls?" "Right. I'll put the honey in. I like watching it drip off the spoon." Something about her tone has caught my attention: the modulation of her voice, its unpressured directness—measured, and with a warmth I have not heard in her in months. Her eyes have softened. I caution myself not to be fooled. Yet the change in her is unmistakable. . . . It's as if a miracle has occurred. The miracle of normalcy, of ordinary existence. . . .
>
> It feels as if we have been living all summer inside a fable. A beautiful girl is turned into a comatose stone or a demon. She is separated from her loved ones, from language, from

everything that had been hers to master. Then the spell is broken and she is awake again.

After her summer of madness, Sally is able to return to school—anxious, but determined to reclaim her life. At first, she keeps her illness to herself and enjoys the company of three close friends from her class. "Often," her father writes, "I listen to her on the phone with them, intimate, biting, gossipy—the buoyant sound of health." A few weeks into the school year, after much discussion with her parents, Sally tells her friends about her psychosis:

> They readily accept the news. Being an alumna of the psych ward confers social status on Sally. It's a kind of credential. She has been where they have not been. It becomes their secret.

Sally's madness resolves, and this, one might hope, would be the end of the story. But the very defining feature of manic-depressive illness is its cyclical nature, and in a post-script to his book, Greenberg indicates that Sally did have two further attacks: four years later, when she was in college, and six years after that (when her medication was discontinued). There is no "cure" for manic-depressive illness, but living with manic-depressive illness may be greatly helped by medication, by insight and understanding (in particular, by minimizing stressors like sleep loss and being alert to the earliest signs of mania or depression), and, not least, by counseling and psychotherapy.

In its detail, depth, richness, and sheer intelligence, *Hurry*

*Down Sunshine* will be recognized as a classic of its kind, along with the memoirs of Kay Redfield Jamison and John Custance. But what makes it unique is the fact that so much here is seen through the eyes of an extraordinarily open and sensitive parent—a father who, while never descending into sentimentality, has remarkable insight into his daughter's thoughts and feelings, and a rare power to find images and metaphors for almost unimaginable states of mind.

The question of "telling," of publishing detailed accounts of patients' lives, their vulnerabilities, their illness, is a matter of great moral delicacy, fraught with pitfalls and perils of every sort. Is Sally's struggle with psychosis not a private and personal matter, no one's business but her own (and that of her family and physicians)? Why would her father consider exposing his daughter's travails, and his family's pain, to the world? And how would Sally feel about a public disclosure of her teenage torments and exaltations?

This was not a quick or easy decision for either Sally or her father. Greenberg did not grab a pen and start writing during his daughter's psychosis in 1996—he waited, he pondered, he let the experience sink deep into him. He had long, searching discussions with Sally, and only more than a decade later did he feel that he might have the balance, the perspective, the tone that *Hurry Down Sunshine* would need. Sally, too, had come to feel this, and urged him not only to write her story but to use her real name, without camouflage. It was a courageous decision, given the stigma and misunderstanding that still surround mental illness of any kind.

It is a stigma that affects many, for manic-depressive illness occurs in all cultures, and affects at least one person in a hundred—there are, at any time, millions of people, some even younger than Sally, who may have to face what she did. Lucid, realistic, compassionate, illuminating, *Hurry Down Sunshine* may provide a sort of guide for those who have to negotiate the dark regions of the soul—a guide, too, for their families and friends, for all those who want to understand what their loved ones are going through.

Perhaps, too, it will remind us of what a narrow ridge of normality we all inhabit, with the abysses of mania and depression yawning to either side.

# The Lost Virtues of the Asylum

W e tend to think of mental hospitals as snake pits, hells of chaos and misery, squalor and brutality. Most of them, now, are shuttered and abandoned—and we think with a shiver of the terror of those who once found themselves confined in such places. So it is salutary to hear the voice of an inmate, one Anna Agnew, judged insane in 1878 (such decisions, in those days, were made by a judge, not a physician) and "put away" in the Indiana Hospital for the Insane. Anna was admitted to the hospital after she made increasingly distraught attempts to kill herself and tried to kill one of her children with laudanum. She felt profound relief when the institution closed protectively around her, and especially from having her madness recognized. As she later wrote:

> Before I had been an inmate of the asylum a week, I felt a greater degree of contentment than I had felt for a year previous. Not that I was reconciled to life, but because my unhappy condition of mind was understood, and I was treated accordingly. Besides, I was surrounded by others in like bewildered, discontented mental states in whose miser-

ies . . . I found myself becoming interested, my sympathies becoming aroused. . . . And at the same time, I too, was treated as an insane woman, a kindness not hitherto shown to me.

Dr. Hester being the first person kind enough to say to me in answer to my question, "Am I insane?" "Yes, madam, and very insane too!" . . . "But," he continued, "we intend to benefit you all we can and our particular hope for you is the restraint of this place." . . . I heard him [say] once, in reprimanding a negligent attendant: "I stand pledged to the State of Indiana to protect these unfortunates. I am the father, son, brother and husband of over three hundred women . . . and I'll see that they are well taken care of!"

Anna also spoke (as Lucy King recounts in her book *From Under the Cloud at Seven Steeples*) of how crucial it was, for the disordered and disturbed, to have the order and predictability of the asylum:

This place reminds me of a great clock, so perfectly regular and smooth are its workings. The system is perfect, our bill of fare is excellent, and varied, as in any well-regulated family. . . . We retire at the ringing of the telephone at eight o'clock, and an hour later, there's darkness and silence . . . all over this vast building.

The old term for a mental hospital was "lunatic asylum," and "asylum," in its original usage, meant refuge, protection, sanctuary—in the words of the *Oxford English Dictionary*, "a benevolent institution according shelter and support to some class of the afflicted, the unfortunate, or destitute." From at least the fourth century A.D., monasteries, nunneries, and churches were places of asylum. And to these were added secular asylums, which (so Michel Foucault suggests) emerged with the virtual annihilation of Europe's lepers by the Black Death and the use of the now-vacant leprosaria to house the poor, the ill, the insane, and the criminal. Erving Goffman, in his famous book *Asylums,* ranks all of these together as "total institutions"—places where there is an unbridgeable gulf between staff and inmates, where rigid rules and roles preclude any sense of fellowship or sympathy, and where inmates are deprived of all autonomy or freedom or dignity or self, reduced to nameless ciphers in the system.

By the 1950s, when Goffman was doing his research at St. Elizabeths Hospital in Washington, D.C., this was indeed the case, at least in many mental hospitals. But creating such a system was hardly the intent of the high-minded citizens and philanthropists who had been moved to found many of America's lunatic asylums in the early and middle years of the nineteenth century. In the absence of specific medications for mental illness at this time, "moral treatment"—a treatment directed towards whole individuals and their potential for physical and mental health, not just a malfunctioning part of their brain—was considered the only humane alternative.

These first state hospitals were often palatial buildings, with high ceilings, lofty windows, and spacious grounds, providing abundant light, space, and fresh air, along with exercise and a varied diet. Most asylums were largely self-supporting and grew or raised most of their own food. Patients would work in the fields and dairies, work being considered a central form of therapy for them, as well as supporting the hospital. Community and companionship, too, were central—indeed, vital—for patients who would otherwise be isolated in their own mental worlds, driven by their obsessions or hallucinations. Also crucial was the recognition and acceptance of their insanity (this, for Anna Agnew, was a great "kindness") by the staff and other inmates around them.

Finally, coming back to the original meaning of "asylum," these hospitals provided control and protection for patients, both from their own (perhaps suicidal or homicidal) impulses and from the ridicule, isolation, aggression, or abuse so often visited upon them in the outside world. Asylums offered a life with its own special protections and limitations, a simplified and narrowed life perhaps, but within this protective structure the freedom to be as mad as one liked and, for some patients at least, to live through their psychoses and emerge from their depths as saner and stabler people.

In general, though, patients remained in asylums for long terms. There was little preparation for a return to life outside, and perhaps after years cloistered in an asylum, residents became "institutionalized" to some extent: they no longer desired or could no longer face the outside world. Patients often lived in state hospitals for decades, and died in them—every asylum had its own graveyard. (Such lives have been reconstructed with great sensitivity by Darby Penney and Peter Stastny in their book *The Lives They Left Behind*.)


IT WAS INEVITABLE, under these circumstances, that the asylum population should grow—and individual asylums, often immense to begin with, came to resemble small towns. Pilgrim

State Hospital, on Long Island, housed nearly fourteen thousand patients at one time. Inevitable, too, that with these huge numbers of inmates, and inadequate funding, state hospitals fell short of their original ideals. By the latter years of the nineteenth century, they had already become known for squalor and negligence, and were often run by inept, corrupt, or sadistic bureaucrats—a situation that persisted through the first half of the twentieth century.

There was a similar evolution, or devolution, at Creedmoor State Hospital in Queens, New York, which had been established in 1912, very modestly, as the Farm Colony of Brooklyn State Hospital, holding to the nineteenth-century ideals of providing space, fresh air, and farming for its patients. But Creedmoor's population soared—it reached seven thousand by 1959—and, as Susan Sheehan showed in her 1982 book, *Is There No Place on Earth for Me?*, it became, in many ways, as wretched, overcrowded, and understaffed as any other state hospital. Yet the original gardens and livestock were maintained, providing a crucial resource for some patients, who could care for animals and plants, even though they might be too disturbed or too ambivalent to maintain relationships with other human beings.

At Creedmoor there were gymnasiums, a swimming pool, and recreation rooms with Ping-Pong and billiards tables; there was a theater and a television studio, where patients could produce, direct, and act in their own plays—plays that, like de Sade's theater in the eighteenth century, could allow creative expression of the patients' own concerns and predicaments. Music was important—there was a small patients' orchestra—and so, too, was visual art. (Even today, with the bulk of the hospital closed down and falling into decay, the remarkable Living Museum at

Creedmoor provides patients with the materials and space to work on painting and sculpture. One of the Living Museum's founders, Janos Marton, calls it a "protected space" for the artists.)

There were gigantic kitchens and laundries, and these, like the gardens and livestock, provided work and "work therapy" for many of the patients, along with opportunities for learning some of the skills of daily life, which, with their withdrawal into mental illness, they might never have otherwise acquired. And there were great communal dining rooms, which, at their best, fostered a sense of community and companionship.

Thus, even in the 1950s, when conditions in state hospitals were so dismal, some of the good aspects of an asylum life were still to be found in them. There were often, even in the worst hospitals, pockets of human decency, of real life and kindness.

The 1950s brought the advent of specific antipsychotic drugs, drugs that seemed to promise at least some alleviation or suppression of psychotic symptoms, if not a "cure" for them. The availability of these drugs strengthened the idea that hospitalization need not be custodial or lifelong. If a short stay in the hospital could "break" a psychosis and be followed by patients returning to their own communities, where they could be maintained on medication and monitored in outpatient clinics, then, it was felt, the prognosis, the whole natural history of mental illness, might be transformed, and the vast and hopeless population of asylums drastically reduced.

DURING THE 1960S, a number of new state hospitals dedicated to short-term admissions were built on this premise. Among these was Bronx State Hospital (now Bronx Psychiatric Center). Bronx State had a gifted and visionary director and a handpicked staff when it opened in 1963, but for all its forward-looking orientation, it had to deal with an enormous influx of patients from the older hospitals, which were now starting to be closed down. I began work as a neurologist there in 1966, and over the years, I was to see hundreds of such patients, many of whom had spent most of their adult lives in hospitals.

There were, at Bronx State as at all such hospitals, great variations in the quality of patient care: there were good, sometimes exemplary, wards, with decent, thoughtful physicians and attendants, and bad, even hideous ones, marked by negligence and cruelty. I saw both of these in my twenty-five years at Bronx State. But I also have memories of how some patients, no longer violently psychotic or on locked wards, might wander tranquilly around the grounds, or play baseball, or go to concerts or films. Like the patients at Creedmoor, they could produce shows of their own, and at any time, patients could be found reading quietly in the hospital library or looking at newspapers or magazines in the dayrooms.

Sadly and ironically, soon after I arrived in the 1960s, work opportunities for patients virtually disappeared, under the guise of protecting their rights. It was considered that having patients work in the kitchen or laundry or garden, or in sheltered workshops, constituted "exploitation." This outlawing of work—based on legalistic notions of patients' rights and not on their real needs—deprived many patients of an important form of therapy, something that could give them incentives and identities of an economic and social sort. Work could "normalize" and create community, could take patients out of their solipsistic inner worlds, and the effects of stopping it were demoralizing in the extreme. For many patients who had previously enjoyed work and activity, there was now little left but sitting, zombie-like, in front of the now-never-turned-off TV.

THE MOVEMENT FOR DEINSTITUTIONALIZATION, starting as a trickle in the 1960s, became a flood by the 1980s, even though it was clear by then that it was creating as many problems as it solved. The enormous homeless population, the "sidewalk psychotics" in every major city, were stark evidence that no city had an adequate network of psychiatric clinics and halfway houses, or the infrastructure to deal with the hundreds of thousands of patients who had been turned away from the remaining state hospitals.

The antipsychotic medications that had ushered in this wave of deinstitutionalization often turned out to be much less miraculous than originally hoped. They might lessen the "positive" symptoms of mental illness—the hallucinations and delusions of schizophrenia. But they did little for the "negative" symptoms—the apathy and passivity, the lack of motivation and ability to relate to others—which were often more disabling than the positive symptoms. Indeed (at least in the manner they were originally used), the antipsychotic drugs tended to lower energy and vitality and produce an apathy of their own. Sometimes there were intolerable side effects, movement disorders like parkinsonism or tardive dyskinesia, which could persist for years after the medication had been stopped. And sometimes patients were unwilling to give up their psychoses, psychoses that gave meaning to their worlds and situated them at the center of these worlds. So it was common for patients to stop taking the anti-psychotic medicine they had been prescribed.

Thus many patients who were given antipsychotic drugs and discharged had to be readmitted weeks or months later. I saw scores of such patients, many of whom said to me, in effect, "Bronx State is no picnic, but it is infinitely better than starving, freezing on the streets, or being knifed on the Bowery." The hospital, if nothing else, offered protection and safety—offered, in a word, asylum.

By 1990 it was very clear that the system had overreacted, that the wholesale closings of state hospitals had proceeded far too rapidly, without any adequate alternatives in place. It was not wholesale closure that the state hospitals needed but fixing: dealing with the overcrowding, the understaffing, the negligences and brutalities. For the chemical approach, while necessary, was not enough. We forgot the benign aspects of asylums, or perhaps we felt we could no longer afford to pay for them: the spaciousness and sense of community, the place for work and play, and for the gradual learning of social and vocational skills—a safe haven that state hospitals were well equipped to provide.

ONE MUST NOT BE too romantic about madness, or the mad-houses in which the insane were confined. There is, under the manias and grandiosities and fantasies and hallucinations, an immeasurably deep sadness about mental illness, a sadness that is reflected in the often grandiose but melancholy architecture of the old state hospitals. As the photographs in Christopher Payne's book *Asylum* attest, their ruins, desolate today in a different way, offer a mute and heartbreaking testimony both to the pain of those with severe mental illness and to the once-heroic structures that were built to try to assuage that pain.

Payne is a visual poet as well as an architect by training, and he spent years finding and photographing these buildings—often the pride of their local communities and a powerful symbol of humane caring for those less fortunate. His photographs are beautiful images in their own right, and they also pay tribute to a sort of public architecture that no longer exists. They focus on the monumental and the mundane, the grand façades and the peeling paint.

Payne's photographs are powerfully elegiac, perhaps especially so for someone who has worked and lived in such places and seen them full of people, full of life. The desolate spaces evoke the lives that once filled them, so that, in our imaginations, the empty dining rooms are once more thronged with people, and the spacious dayrooms with their high windows again contain, as they once did, patients quietly reading or sleeping on sofas or (as was perfectly permissible) just staring into space. They evoke for me not only the tumultuous life of such places but the protected and special atmosphere they offered when, as Anna Agnew noted in her diary, they were places where one could be both mad and safe, places where one's madness could be assured of finding, if not a cure, at least recognition and respect, and a vital sense of companionship and community.


WHAT IS THE SITUATION NOW? The state hospitals that still exist are almost empty and contain only a tiny fraction of the numbers they once had. The remaining inmates consist for the most part of chronically ill patients who do not respond to medication or incorrigibly violent patients who cannot be safely allowed outside. The vast majority of mentally ill people therefore live outside mental hospitals. Some live alone or with their families and visit outpatient clinics, and some stay in "halfway houses," residences that provide a room, one or more meals, and the medications that have been prescribed.

Such residences vary greatly in quality—but even in the best of them (as brought out by Tim Parks in his review of Jay Neugeboren's book about his schizophrenic brother, *Imagining Robert,* and by Neugeboren himself, in his recent review of *The Center Cannot Hold,* Elyn Saks's autobiographical account of her own schizophrenia), patients may feel isolated and, worst of all, scarcely able to get the psychiatric advice and counseling they may need.[1] The last few decades have seen a new generation of antipsychotic drugs, with better therapeutic effects and fewer side effects, but the too-exclusive emphasis on chemical models of schizophrenia, and on purely pharmacological approaches to treatment, may leave the central human and social experience of being mentally ill untouched.

Particularly important in New York City—especially since deinstitutionalization—is Fountain House. Established in 1948, it provides a "clubhouse" on West Forty-Seventh Street for mentally ill people from all over the city. Here they can come and go freely, meet others, eat communally, and, most importantly, use resources and networks for finding jobs or apartments, furthering their education, navigating the health care system, and so on. Similar clubhouses have now been established in many

1. Elyn Saks, who has lived with schizophrenia since childhood, is a MacArthur Foundation fellow and a professor at the USC Gould Law School, where she specializes in mental health and the law.

cities. There are dedicated staff members and volunteers at these clubhouses, but they are crucially dependent on private funds, since public funding is very inadequate.

ANOTHER MODEL EXISTS, in the little Flemish town of Geel, not far from Antwerp. Geel is a unique social experiment—if one can use the word "experiment" for something that has gone on for seven centuries and arose in so natural and spontaneous a fashion. In the seventh century, legend has it, Dymphna, the daughter of an Irish king, fled to Geel to avoid the incestuous embrace of her father, and he, in a murderous rage, had her beheaded. She was worshipped in Geel as the patron saint of the mad, and her shrine soon attracted mentally ill people from all over Europe. By the thirteenth century, the families in this little Flemish town had opened their homes and hearts to the mentally ill—and they have been doing so ever since. For centuries, it was the norm for a family in Geel to take in or adopt a boarder; during more agricultural times, these "guests" would be a welcome source of labor.

The tradition is waning today, even though such families now receive a modest government subsidy. But when a family—often a couple with young children—indicates its readiness to take in a guest, they make no inquiry about their guest's psychiatric condition or diagnosis. The guests are brought into the home as individuals, and when the relationship works well, as it does most of the time, the guests become cherished family members, like a beloved aunt or uncle. They may play a part in bringing up the children and grandchildren or in taking care of the elders.

The anthropologist Eugeen Roosens has studied Geel in

depth for over thirty years; he first published his observations in 1979 (*Mental Patients in Town Life: Geel—Europe's First Therapeutic Community*). As he and his colleague Lieve Van de Walle write, the Geel solution is "not simply a happy but isolated remnant of the Middle Ages." The system there has undergone at least two fundamental transformations that have allowed it to remain viable. The first occurred when the Belgian government introduced medical supervision in the community and, in 1861, built a hospital. Here, if things got too difficult for the family to handle, a boarder could go for medical treatment. Thus fortified by a hospital and its professional staff—psychiatrists, nurses, social workers, and therapists—providing both help for the family and (if need be) medical care, Geel continued to flourish, at one time before World War II hosting several thousand mentally ill boarders.

A second change, over the last fifty years, has occurred as the impact of health professionals in Geel has markedly increased. During the day, more than half of the patients are able to take part in jobs or day programs, away from home, supervised by therapists and social workers. (Intentionally or not, this rise in day-care situations has coincided with a decline in home-based work, as more and more of the host families have moved into nonagricultural jobs.)

Thus, Geel has evolved into a two-layered system, but a number of key elements of the traditional system have remained intact. Chief among them, as Roosens and Van de Walle write, are "maximal kin-like inclusion and integration of the patient, the kindness of the broader social context in Geel, the acceptance of the patients' inherent limitations, the strong bond between boarders and foster families, the resilient mutual loyalty, and the entrenched responsibility of the next generation towards the boarders."[2]

When I visited the town a few years ago, I saw guests strolling or bicycling in the streets, chatting, working in shops. I would not have guessed that they were boarders (save for clues from occasional odd mannerisms or behaviors), except that my hosts, from the hospital, knew all of them individually and were able to identify them for me. In the world at large, the mentally ill are often isolated, stigmatized, shunned, feared, seen as less than fully human. But here, in this little town, they were respected as fellow beings, treated with affection and care—at least as much as anyone else.

When I asked several foster families why they had welcomed such a guest, they seemed confused. Why would they not? Their parents and grandparents had all done the same; here it is a way of life. The people of Geel may know that their neighbor, say,

---

2. Roosens and Van de Walle are themselves members of this community, part of the fabric of life in Geel. They are thus able to provide nineteen detailed portraits of families and their boarders, some of whom Roosens has observed over a span of decades. These families and their guests present a wide range of situations, from happy ones in which the hosts and guests love and care for each other deeply, to homes in which the guests are "difficult" (Geelians speak of "good" boarders or, more rarely, "difficult" ones but never of "bad" or "crazy" boarders)—so difficult that the fostering situation has broken down. Even where very severe psychiatric problems exist, Roosens points out, when "a mutual warm relationship evolves [as it usually does], foster parents are prepared to go to great lengths to accommodate their guests."

These nineteen case studies are exemplary in their richness and detail, and constitute primary material of enormous value. Along with the rest of the book, they provide a definitive rebuttal of the notion of mental illness as a remorselessly advancing and deteriorating condition and show how, if there can be an effective integration into family and community life (and, behind this, a safety net of hospital care, professionals, and medication where warranted), even those who would seem to be incurably afflicted can, potentially, live full, dignified, loved, and secure lives.

is a guest, with mental problems of one sort or another—but such a fact lends, seemingly, no stigma. It is simply a fact of life, unremarkable, just as being male or female would be.

"For the residents of Geel," Roosens and Van de Walle write,

> the line between "patients" and ordinary people is, in many ways, nonexistent. The prejudices against mental illness that are very much alive in the world at large are not found in the people of Geel, because they have been raised, for many generations, in the presence of mental "patients." What makes Geel remarkable is not the blurring of the boundary between normal and abnormal, but the recognition of each patient's human dignity, to the extent that, for them, family and community life is given an honest chance every single day.

At the beginning of the nineteenth century, when Philippe Pinel, a founder of psychiatry in France, appealed to the new revolutionary government to strike off the chains that had (often literally) been used for centuries to shackle mad people, and a humane breath swept over Europe, this question was epitomized by Geel. Could a place like Geel provide a real alternative?

WHILE GEEL IS UNIQUE, there are other residential communities that derive, historically, both from the asylums and the therapeutic farm communities of the nineteenth century, and these provide, for the fortunate few who can go to them, comprehensive programs for the mentally ill. I have visited some of these—including Gould Farm in the Berkshires and CooperRiis near Asheville, North Carolina—and have seen in them much of what was admirable in the life of the old state hospitals. In places like these, Goffman's gulf between staff and inmates is almost erased. There are friendships, and there is work for everyone to do. The cows have to be milked, the corn harvested. At the communal dinners at Gould Farm, I often found it impossible to know who was staff and who was a resident. Residents regularly move on to becoming staff. Community, companionship, opportunities for work and creativity, and respect for the individuality of everyone there—these things are coupled with psychotherapy and whatever medication is needed.

Often the medication is rather modest in these ideal circumstances. Many of the patients in such places (though schizophrenia and manic depression remain lifelong conditions) may graduate after several months or perhaps a year or two, moving into more independent living, and perhaps going back to work or school, with a more modest degree of ongoing support and counseling. For many of them, a full and satisfying life with fewer or even no relapses is within reach.

Although the cost of such residential facilities is considerable—more than a hundred thousand dollars a year (some of which is funded by family contributions, the rest by private donors)—this is far less than the cost of a year in a hospital, to say nothing of the human costs involved. But there are only a handful of comparable facilities in the United States—they can accommodate no more than a few hundred patients.

The remainder—the 99 percent of the mentally ill who have insufficient resources of their own—must face inadequate treatment and lives that cannot reach their potential. The millions of mentally ill remain the least supported, the most disenfranchised, and the most excluded people in our society today. And yet it is clear—from the experiences of places like CooperRiis and Gould Farm, and of individuals like Elyn Saks—that schizophrenia and other mental illnesses are not necessarily relentlessly deteriorating (although they can be); and that, in ideal circumstances, and when resources are available, even the most deeply ill people—those who have been relegated to a "hopeless" prognosis—may be enabled to live satisfying and productive lives.

# Life Continues

# Anybody Out There?

One of the first books I read as a boy was H. G. Wells's 1901 fable, *The First Men in the Moon.* The two men, Cavor and Bedford, land in an apparently barren and lifeless crater just before the lunar dawn. Then, as the sun rises, they realize there is an atmosphere—they spot small pools and eddies of water, and then little round objects scattered on the ground. One of these, as it is warmed by the sun, bursts open and reveals a sliver of green. "A seed," says Cavor, and then, very softly, "Life!" They light a piece of paper and throw it onto the surface of the moon. It glows and sends up a thread of smoke, indicating that the atmosphere, though thin, is rich in oxygen and will support life as they know it.

This was how Wells conceived the prerequisites of life: water, sunlight (a source of energy), and oxygen. "A Lunar Morning," the eighth chapter in his book, was my first introduction to astrobiology.[1]

---

1. If Wells envisaged the beginning of life in *The First Men in the Moon,* he envisaged its ending in *The War of the Worlds,* where the Martians, confronting increasing desiccation and loss of atmosphere on their own planet, make a desperate bid to take over the Earth (only to perish from infection by terrestrial bacteria). Wells, who had trained as a biologist, was very aware of both the toughness and the vulnerability of life.

It was apparent, even in Wells's day, that most of the planets in our solar system were not possible homes for life. The only reasonable surrogate for the Earth was Mars, which was known to be a solid planet of reasonable size, in stable orbit, not too distant from the sun, and so, it seemed likely, having a range of surface temperatures compatible with the presence of liquid water.

But free oxygen gas—how could that occur in a planet's atmosphere? What would keep it from being mopped up by ferrous iron and other oxygen-hungry chemicals on the surface unless, somehow, it was continuously pumped out in huge quantities, enough to oxidize all the surface minerals and keep the atmosphere charged as well?

It was the blue-green algae, or cyanobacteria, that must have infused the Earth's atmosphere with oxygen, a process that took more than a billion years. The cyanobacteria invented photosynthesis: by capturing the energy of the sun, they were able to combine carbon dioxide (massively present in the Earth's early atmosphere) with water to create complex molecules—sugars, carbohydrates—which the bacteria could then store and tap for energy as needed. This process generated free oxygen as a by-product, a waste product that was to determine the future course of evolution.

Although free oxygen in a planet's atmosphere would be an infallible marker of life, and one that, if present, should be readily detected in the spectra of extrasolar planets, it is not a prerequisite for life. Planets, after all, get started without free oxygen, and may remain without it all their lives. Anaerobic organisms swarmed before oxygen was available, perfectly at home in the atmosphere of the early Earth, converting nitrogen to ammonia, sulfur to hydrogen sulfide, carbon dioxide to formaldehyde, and so forth. (From formaldehyde and ammonia the bacteria could make every organic compound they needed.)

There may be planets in our solar system and elsewhere that lack an atmosphere of oxygen but are nonetheless teeming with anaerobes. And such anaerobes need not live on the surface of the planet; they could occur well below the surface, in boiling vents and sulfurous hot pots, as they do on Earth today, to say nothing of subterranean oceans and lakes. (There is thought to be such a subsurface ocean on Jupiter's moon Europa, locked beneath a shell of ice several miles thick, and its exploration is one of the astrobiological priorities of this century. Curiously, Wells, in *The First Men in the Moon,* imagines life originating in a central sea in the middle of the moon and then spreading outward to its inhospitable periphery.)

IT IS NOT CLEAR whether life has to "advance," whether evolution must take place, if there is a satisfactory status quo. Brachiopods, lampshells, for instance, have remained virtually unchanged since they first appeared in the Cambrian period, more than five hundred million years ago. But there does seem to be a drive for organisms to become more highly organized and more efficient in retaining energy, at least when environmental conditions are changing rapidly, as they were before the Cambrian. The evidence indicates that the first primitive anaerobes on Earth were prokaryotes: small, simple cells—just cytoplasm, usually bounded by a cell wall, but with little if any internal structure.

Primitive as they are, prokaryotes are still highly sophisticated organisms, with formidable genetic and metabolic machinery. Even the simplest ones manufacture more than five hundred proteins, and their DNA includes at least half a million base pairs. Certainly still more primitive life-forms must have preceded the prokaryotes.

Perhaps, as the physicist Freeman Dyson has suggested, there were progenotes capable of metabolizing, growing, and dividing but lacking any genetic mechanism for precise replication. And before them there must have been millions of years of purely chemical, prebiotic evolution—the synthesis, over eons, of formaldehyde and cyanide, of amino acids and peptides, of proteins and self-replicating molecules. Perhaps that chemistry took place in the minute vesicles, or globules, that develop when fluids at very different temperatures meet, as may well have happened around the boiling midocean vents of the Archean sea.

By degrees, however—and the process took place with glacial slowness—prokaryotes became more complex, acquiring internal structure, nuclei, mitochondria, and so on. The microbiologist Lynn Margulis has suggested that these complex so-called eukaryotes arose when prokaryotes began incorporating other prokaryotes within their own cells. The incorporated organisms at first became symbiotic and later came to function as essential organelles of their hosts, enabling the resultant organisms to utilize what was originally a noxious poison: oxygen.

THE TWO PREEMINENT evolutionary changes in the early history of life on Earth—from prokaryote to eukaryote, from anaerobe to aerobe—took the better part of two billion years.

And then another thousand million years had to pass before life rose above the microscopic and the first multicellular organisms appeared. So if the Earth's history is anything to go by, one should not expect to find any higher life on a planet that is still young. Even if life has appeared and all goes well, it could take billions of years for evolutionary processes to move it along to the multicellular stage.

Moreover, all those "stages" of evolution—including the evolution of intelligent, conscious beings from the first multicellular forms—may have happened against daunting odds, as Stephen Jay Gould and Richard Dawkins, in their different ways, have brought out. Gould spoke of life as "a glorious accident"; Dawkins likens evolution to "climbing Mount Improbable." And life, once started, is subject to vicissitudes of all kinds: from meteors and volcanic eruptions to global overheating and cooling; from dead ends in evolution to mysterious mass extinctions; and finally (if things get that far) from the fateful proclivities of a species like ourselves.

There are microfossils in some of the Earth's most ancient rocks, rocks more than three and a half billion years old. So life must have appeared within one or two hundred million years after the Earth had cooled off sufficiently for water to become liquid. That astonishingly rapid transformation makes one think that life may develop readily, perhaps inevitably, given the right physical and chemical conditions.

But can one speak confidently of "Earthlike" planets, or is the Earth physically, chemically, and geologically unique? And even if there are other "habitable" planets, what are the chances that life, with its thousands of physical and chemical coincidences and contingencies, will emerge?

Opinion here varies as widely as it can. The biochemist Jacques Monod regarded life as a fantastically improbable accident, unlikely to have arisen anywhere else in the universe. In his book *Chance and Necessity,* he writes, "The universe was not pregnant with life." Another biochemist, Christian de Duve, takes issue with this; he sees the origin of life as determined by a large number of steps, most of which had a "high likelihood of taking place under the prevailing conditions." Indeed, de Duve believes that there is not merely unicellular life throughout the universe but complex, intelligent life, too, on trillions of planets. How are we to align ourselves between these utterly opposite but theoretically defensible positions?

Indeed, life on Earth may have originated elsewhere. We know, from the samples returned by the Apollo missions, that there are early Earth and Martian meteors on the moon in considerable quantities. There must be thousands of Martian meteorites on Earth. The notion of "seed-bearing meteoritic stones" was raised by Lord Kelvin in 1871, and the notion of free spores drifting through space and seeding life on other planets ("panspermia") was postulated by the Swedish chemist Svante Arrhenius a few years later (an idea revived in the twentieth century by Francis Crick and Leslie Orgel, as well as Fred Hoyle). The idea was considered implausible for more than a century, but is once again a subject for discussion. For now it is evident that the insides of sizable meteors do not get heated to sterilizing temperatures, and that bacterial spores or other resistant forms could, in principle, survive within them, protected by the body of the meteor not only from heat but from radiations deadly to life. Meteors were being flung in all directions during the period of Heavy Bombardment four million years ago. Chunks of the

Earth must have been ejected into space then, as well as chunks of Mars and Venus—a Mars and Venus that might, at the time, have been more hospitable to life than Earth itself.

What we need, what we must have, is hard evidence of life on another planet or heavenly body. Mars is the obvious candidate: it was wet and warm there once, with lakes and hydrothermal vents and perhaps deposits of clay and iron ore. It is especially in such places that we should look, and if the evidence shows that life once existed on Mars, we will then need to know, crucially, whether it originated there or was transported (as would have been readily possible) from the young, teeming, volcanic Earth. If we can determine that life originated independently on Mars (if Mars, for instance, once harbored DNA nucleotides different from our own), we will have made an incredible discovery—one that will alter our view of the universe and enable us to perceive it, in the words of the physicist Paul Davies, as a "bio-friendly" one. It would help us to gauge the probability of finding life elsewhere instead of bombinating in a vacuum of data, caught between the poles of inevitability and uniqueness.

IN JUST THE PAST FEW DECADES, life has been discovered in previously unexpected places on our own planet, such as the life-rich black smokers of the ocean depths, where organisms thrive in conditions biologists would once have dismissed as utterly deadly. Life is much tougher, much more resilient, than we once thought. It now seems to me quite possible that micro-organisms or their remains will be found on Mars, and perhaps on some of the satellites of Jupiter and Saturn.

It seems far less likely—many orders of magnitude less

likely—that we will find any evidence of higher-order, intelligent life-forms, at least in our own solar system. But who knows? Given the vastness and the age of the universe at large, the innumerable stars and planets it must contain, and our radical uncertainties about life's origin and evolution, the possibility cannot be ruled out. And though the rate of evolutionary and geochemical processes is incredibly slow, that of technological progress is incredibly fast. Who is to say (if humanity survives) what we may not be capable of, or discover, in the next thousand years?

For myself, since I cannot wait, I turn to science fiction on occasion—and, not least, back to my favorite Wells. Although it was written a hundred years ago, "A Lunar Morning" has the freshness of a new dawn, and it remains for me, as when I first read it, the most poetic evocation of how it may be when, finally, we encounter alien life.

# Clupeophilia

Anyone on the sixteenth floor of the Roger Smith Hotel in midtown at 5:45 on a recent June afternoon would have seen a puzzling assemblage of people in the corridor: a construction worker from Brooklyn, a mathematics professor from Princeton, a couple from Aruba, a father with an infant strapped to his chest, and an artist from the Lower East Side. It wasn't immediately apparent what had brought this seemingly random slice of humanity together. Had one come up in the service elevator, though, an unmistakable aroma would have given a vital clue. By 5:59, almost sixty people had gathered in the hallway.

At six, the doors to an event room opened and the crowd rushed in. There, in the middle of the room, lighted, draped, surmounted by a huge glittering block of ice, was an altar: an altar covered with hundreds of fresh herring, the first of the season, just flown in from Holland. This was an altar consecrated to Clupeus, the god of herring, whose annual festival is celebrated in late spring by herring lovers the world over.

Entire books have been published about cod, about eel,

about tuna, but relatively little has been written about herring. (There is, however, a delightful book by Mike Smylie, *Herring: A History of the Silver Darlings,* and a fascinating chapter in W. G. Sebald's *The Rings of Saturn.*) But herring have played a great part in human history. In the Middle Ages, they were carefully graded and priced by the Hanseatic League, and supported fisheries in the Baltic and the North Sea—and, later, in Newfoundland and on the Pacific Coast. Herring are one of the commonest, cheapest, and most delicious fish on the planet—a fish that can take an infinity of forms: marinated, pickled, salted, fermented, smoked, or, as with the exquisite Hollandse Nieuwe, straight from the sea. They are one of the healthiest fish, too, full of omega-3 oils, and without the mercury that accumulates in the big predators such as tuna and swordfish. A few years ago, the oldest person in the world, a 114-year-old Dutch woman, said she attributed her longevity to eating pickled herring every day. (A 114-year-old woman from Texas attributed her long life to "minding my own business.")

There are many species of Clupeidae, with varying sizes and tastes, from the Atlantic herring, *Clupea harengus,* to the pilchard (much loved in England, and often served in tomato sauce), to the tiny sprat, best smoked and eaten bones and all. When I grew up in England, in the 1930s, we had herring virtually every day: smoked herring (kippers or bloaters) at breakfast, perhaps a herring pie at lunch (my mother's favorite dish), fried herring roe on toast at teatime, chopped herring at dinner. But times have changed, herring is no longer on every breakfast and dinner table, and it is only on special, joyous occasions that we clupeophiles can come together for a real herring feast.

The great traditions of herring are maintained by Russ &
Daughters, a Houston Street emporium that started as a push-
cart on the Lower East Side over a century ago and still sells
the largest variety of herring in New York City. It was Russ &
Daughters that organized the recent herring festival.

There are certain passions—one wants to call them innocent, ingenuous passions—that are great democratizers. Baseball, music, and bird-watching come immediately to mind. At
the herring festival, there was no talk about the stock market, no
gossiping about celebrities. People had come to eat herring—to
savor them, to compare them. In its purest form, this meant seiz-
ing the new herring by the tail and lowering them gently into the
mouth. The sensation this produces is voluptuous, especially as
they slip down the throat.

Guests started from the great central table, the altar covered
with new herring; washed these down with aquavit; and moved
on to satellite tables, where there were *matjes* herring, herring
in wine sauce, herring in cream sauce, Bismarck herring, herring
in mustard sauce, herring in curry sauce, and plump schmaltz
herring, fresh from Iceland. Oily and briny, schmaltz herring can
last for twenty years; taken from the Baltic, they were a staple
food (along with black bread, potatoes, and cabbage) of poor
Jews throughout Eastern Europe. For my father, born in Lithuania, there was nothing to compare with them, and he ate them
daily all his life.

Around eight o'clock, after two hours of eating and drink-
ing, the pace slackened. Slowly, the herring lovers left the hotel,
still discussing favorite dishes with fellow travelers as they went.
They sauntered slowly up Lexington Avenue. One does not rush

after such a banquet; indeed, one's whole perspective on the world is changed. Some of us, the New Yorkers, will meet again, at Russ & Daughters. But the rest, after they have slept the deep sleep of the consummated herring eater, will start counting the days to next year's herring festival.

# Colorado Springs Revisited

The limo driver who has picked me up at the Colorado Springs airport is taking me to the Broadmoor—I know nothing about it, but he pronounces the name with a sort of reverence or awe—says, "You've stayed there before?"

No, I say, the last time I was in Colorado Springs was in 1960, and then I was zigzagging around the country on my motorbike, with a bedroll on the back. He digests this. "Real swank place, the Broadmoor," he says at length.

It is indeed—all three thousand acres of it—a sort of Hearst Castle, with a lake, three golf courses, fake four-posters in the bedrooms, and flunkies, charming men and women trained to anticipate your every wish and action, pulling out chairs, opening doors, offering suggestions for dinner. How far, I wondered, would this overservice go? Would one of these pleasant, uniformed helpers thrust a tissue under my nose if they saw me about to sneeze? I am uncomfortable being so waited upon, and would prefer to go about my business quietly, open my own doors, pull out my own chairs, blow my own nose.

Later, I am sitting outside on the terrace of one of the Broadmoor's many restaurants, an informal one that just serves, I am told, "simple" bar food. As I sit, gazing at snowcapped Cheyenne Mountain and the beautiful, clear mountain skies, eating a chicken sandwich the size of my head, a plane climbs almost vertically in front of me, leaving shining twin contrails in its wake. I wonder if it is from the U.S. Air Force Academy nearby—no civilian plane, surely, could climb like that—and my mind goes back to 1960–61, when I was biking around the country and paid a special visit to the academy's new chapel, which, with its dramatic, triangular outline, looked as though it were shooting up into the sky.

I was twenty-seven. I had arrived in North America a few months before and started out by hitchhiking across Canada, then down to California, which I had been in love with since I was a fifteen-year-old schoolboy in postwar London. California stood for John Muir, Muir Woods, Death Valley, Yosemite, the soaring landscapes of Ansel Adams, the lyrical paintings of Albert Bierstadt. It meant marine biology, Monterey, and "Doc," the romantic marine biologist figure in Steinbeck's *Cannery Row.*

It was not just physical spaciousness that America stood for in my mind then but moral openness and spaciousness, too. In England, one was classified—working class, middle class, upper class—as soon as one opened one's mouth; one did not mix, one was not at ease, with people of a different class. The system, though implicit, was nonetheless as rigid, as uncrossable, as the caste system in India. America, I imagined, was a classless society, a place where everyone, irrespective of birth, color, religion, education, or profession, could meet each other as fellow humans, brother animals, a place where a professor could talk to a truck driver without the categories coming between them.

I had had a taste, a glimpse, of such a democracy, an equality, when I roved about England on my motorcycle in the 1950s. Even in stiff England, motorcycles seemed to bypass the barriers, to open a sort of social ease and good nature in everyone. "That's a nice bike," someone would say, and the conversation would go from there. I had seen this as a boy when my father had a motorbike (with a sidecar, in which he would take me along), and I encountered it again when I got my own bike. Motorcyclists were a friendly group; we waved to one another when we passed on the road, made conversation easily if we met at a café. We formed a sort of romantic, classless society within society at large.

I ARRIVED IN SAN FRANCISCO in 1960 with only a temporary visa and owning almost nothing except the clothes I wore. I had eight months to wait before I could get a green card and start my internship in a San Francisco hospital, and in that time I wanted to see the whole country—in the most vivid, unshielded, direct way possible—and the way to do that, to my mind, was by motorbike. I borrowed some money, bought an old BMW, and set out with nothing but a bedroll and half a dozen blank notebooks, to encounter the vastness of America. Setting out on Route 66, I biked through California, Arizona, Colorado . . . and this was how I found myself, early in 1961, outside the Air Force Academy.

The academy itself was full of young, idealistic cadets, heroes all, to my impressionable eyes. I had volunteered, a few months before, for the Royal Canadian Air Force—but the RCAF wanted me as a research physiologist, and I wanted to fly. Flying was still invested for me with a sort of glamour. Airmen, to my mind,

were the motorbikers of the air, with goggles and leather hel-
mets and thick leather flying jackets, enjoying ecstasies, facing
dangers, like Saint-Exupéry (and perhaps fated, like him, to die
young).

So I identified with the young cadets—their youth, their aspi-
rations, their optimism, their idealism. It was part and parcel of
my pristine vision of America, that first enchanted encounter with
it, when I was still in love with an America I had dreamed about:
an America of vast spaces and mountains and canyons—young,
innocent, ingenuous, strong, open, as Europe had long ceased
to be—and, by happy coincidence, with a great young president
at its helm.

I was soon to be disenchanted, disillusioned on many fronts.
The death of Kennedy added an almost personal pain. But that
day in the spring of 1961, when I was twenty-seven and full of
vigor and hope and optimism myself—that day, that vision of
Colorado Springs and the Air Force Academy, made my heart
exalt, beat strongly with joy and pride.

This comes back to me with a sense of the ludicrous now (but
one must not condescend to one's younger self), as I sit here in
this plush, false Eden, forty-three years later. I stir slightly in my
chair, and the waiter, telepathic, brings me another beer.

# Botanists on Park

There is no end to the odd things that New Yorkers do on Saturday mornings. This, at least, is what drivers must have thought recently when they had to slow down to avoid a line of a dozen people flattened against the enormous embankment of the Park Avenue railroad trestle, peering with magnifying glasses and monoculars into tiny crevices in the stone. Passersby stared, asked questions, and even took photographs. Police officers stopped their patrol car and watched with suspicion or with bewilderment—until they caught sight of the T-shirts many of us were wearing, which bore slogans such as "American Fern Society" or "Ferns Are Ferntastic." We were assembled for a meeting of the American Fern Society, which had joined with the Torrey Botanical Society for a Saturday morning Fern Foray. These forays, which have been going on for more than a century, are usually in somewhat more bucolic sites, but this time we had no goal beyond the Park Avenue viaduct, which, with its crevices and crumbling mortar, is a perfect place to see chink-finding, xerophytic ferns—ferns that, unlike most, can stand long periods of drying out and come to life again after a good rain.

The AFS is a society of amateurs that was founded in Victorian times, an age of amateurs and naturalists. Darwin is our icon. We include a poet, two schoolteachers, a garage mechanic, a neurologist, a urologist, and assorted others. We are about equally divided by gender, and our ages vary from twenty to eighty. Besides us pteridophiles that morning, there was a young couple, two bryophiles, from the Torrey Botanical Society—a group of botanists and amateurs founded in the 1860s, just a few years before the AFS. They were "slumming" amongst the fern people: their interests lie more in mosses, liverworts, and lichens. Ferns are a bit too modern, too evolutionarily advanced, for them, just as flowering plants are for the rest of us.

One tends to think of ferns as delicate and moisture-loving, and many of them are. But others are among the toughest plants on the planet. Ferns will invariably be the first things to sprout, say, on a new lava flow. The planet's atmosphere is full of fern spores. *Woodsia obtusa,* the basic fern on the Park Avenue embankment, has sixty-four spores within each sporangium, and thousands of sporangia on the underside of the fronds of every plant, so each plant is good for a million spores, perhaps more. Let one of these land on a suitable place, and you see why ferns are the great opportunists of the plant world. Indeed, in the fossil record, there is something called the "fern spike," which shows how, after most of the world's plants and land animals were killed in the great extinction of the late Cretaceous period, life came bursting back in the form of ferns.

The leaders that morning were Michael Sundue, a young botanist and fern expert from the New York Botanical Garden, and Elisabeth Griggs, a botanical illustrator. We started on the west side of the trestle—it is shaded in the morning—and trekked up

Park Avenue, facing the traffic. "Botanize at your own risk," the Fern Foray invitation had said.

"An ideal habitat for gametophytes," Sundue said. "Tiny rivulets of water creep down after a rain, dissolving the mortar, making an ideal medium for the lime-tolerant *Woodsia obtusa.*" He discovered a tiny heart-shaped gametophyte in a bed of moss. It had no fronds and looked nothing like a fern. It much more resembled, the bryological couple was happy to see, a liverwort—but it is a crucial intermediate stage in the fern's reproductive cycle. It has male and female organs on its surface, and when it is fertilized two tiny fronds, the new fern, will sprout from it. On an adult *Woodsia,* Sundue pointed out the tiny black umbrella-shaped structures, the indusia, that shelter the sporangia. When time comes for the sporangia to scatter their spores, they activate an ingenious catapult mechanism that flings the spores into the breeze. The spores will float, perhaps, for miles. And if they land somewhere moist and suitable, they will grow into gametophytes, and the cycle will continue.

High above his head, Sundue spotted a gigantic *Woodsia* specimen, almost six feet across, clinging to the rock. "That one's a good age," he said. "Decades old—some species can be very long-lived." When he was asked if ferns show signs of age, he hesitated; the answer is not clear. A fern tends to keep growing until it outruns its food supply, is ousted by competitors, or (as will happen sooner or later with the *Woodsia*) becomes so heavy that it falls to the ground. In some botanical gardens, there are massive ferns more than a hundred years old. Death is not built in to these plants as it is for us more specialized life-forms, with the ticking clocks of our telomeres, our liability to mutations, our running-down metabolisms. But youth is apparent, even in ferns. The young *Woodsia* are charming: a bright spring green; tiny, like babies' toes; and very soft and vulnerable.

There was nothing but *Woodsia* between 93rd and 104th Streets, but moving to the next block we spotted a *Thelypteris palustris,* the marsh fern, here in a very unmarshy environment. It was perched in the wall about eight feet above the ground. Sundue, acrobatic, leapt up and pulled down a frond. We passed it around, peering at it through high-powered lenses and using Swiss Army knives to dissect its vascular bundles.

One of the group, an angiosperm lady from the Torrey Society, spotted a flowering plant near the *Thelypteris*. It was oozing with sticky white resin. *Lactuca,* she said, allied to lettuce. The word made me think back to my marine biology days, and stimulated a sudden memory of *Ulva lactuca,* the edible seaweed that is often called sea lettuce. I thought, too, of the old word *lactucarium* (which the OED defines as "inspissated juice of various kinds of lettuce, used as a drug").

All these names are irresistible, and the next one seemed positively neurological: ebony spleenwort, *Asplenium platyneuron,* densely covered the trestle between 104th and 105th Streets. It used to be much rarer in this area, Sundue said, but now its range is spreading north and east. Sometimes plants migrate because a favorable habitat has been created. Rocks in New York tend to be acidic, hostile to these alkaline-loving ferns, although artificial structures made with mortar can provide a haven for lime-loving plants. But the great Park Avenue trestle goes back to the nineteenth century, long before the *Asplenium* is believed to have started spreading. Perhaps there is some local source of warmth (cities are full of unexpected heat islands), or perhaps it is yet another sign of global warming—perhaps both.

Between 105th and 106th Streets we found *Onoclea sensibilis,* the "sensitive fern." It looked very dry. It was not doing too well; solicitously, I gave it a drink from my water bottle. If I watered all the *Onocleas* here regularly, Sundue said, they would become the dominant fern species, and completely alter the ecology of the trestle.

Next came another splendidly named fern, *Pellaea atropurpurea.* Some of the plants, those in the densest shade, were a deep blue, almost indigo, verging on purple. None of us were certain why this should be so. Is the blue just a waxy cuticle, or is it a diffraction color like the metallic blue one can see on some butterflies' or birds' wings? Some ferns turn an iridescent blue, in a strategy evolved for absorbing more light. Would this *Pellaea* revert to green in bright light? We gathered some to take home and experiment with in different illuminations.

The block between 109th and 110th Streets was the richest so far. Here—and nowhere else—Griggs's favorite, *Cystopteris tenuis,* grows, along with the remarkable "walking fern," *Asplenium rhizophyllum,* which seems to shoot out new limbs like a brachiating gibbon, putting down suckers at intervals, thus striding across great expanses of stone.

And then, suddenly, strangely, at 110th Street the ferns stopped. From that point north, there was a startling, lifeless desolation, as if someone had decided to eradicate all signs of cryptogamic life. No one knew for sure why this was so, but we quickly crossed over to the sunny side of the trestle and began to work our way south again.

# Greetings from the
# Island of Stability

Early in 2004, the discovery of two new elements—113 and 115—was announced by a team of Russian and American scientists. There is something about such announcements that raises the spirits, thrills one, evokes thoughts of new lands being sighted, of new areas of nature revealed.

It was only at the end of the eighteenth century that the modern idea of an "element" was clearly defined, as a substance that could not be decomposed by any chemical means. In the first decades of the nineteenth century, Humphry Davy, the chemical equivalent of a big-game hunter, thrilled scientists and the public alike by bagging potassium, sodium, calcium, strontium, barium, and a few other elements. Discoveries rolled on throughout the next hundred years, often exciting the public imagination, and when, in the 1890s, five new elements were discovered in the atmosphere, these quickly found their way into H. G. Wells's novels—argon was used by the Martians in *The War of the Worlds,* and helium to make the antigravity material that transported Wells's heroes in *The First Men in the Moon.*

The last naturally occurring element, rhenium, was discovered in 1925. But then, in 1937, there came something no less thrilling: the announcement that a new element had been *created*—an element that seemingly did not exist in nature. The element, number 43, was named "technetium," to emphasize that it was a product of human technology.

It had been thought that there were just ninety-two elements, ending with uranium, whose massive atomic nucleus contained no less than ninety-two protons, along with a considerably larger number of neutral particles (neutrons). But why should this be the end of the line? Could one create elements beyond uranium, even if they did not exist in nature? When, in 1940, Glenn T. Seaborg and his colleagues at the Lawrence Berkeley National Laboratory in California were able to make a new element with ninety-four protons in its huge nucleus, they could not imagine that anything more massive would ever be obtained, and so they called their new element "ultimium" (later it would be renamed plutonium).

If such elements with enormous atomic nuclei did not exist in nature, this was, presumably, because they were too unstable: with more and more protons in the nucleus repelling each other, the nucleus would tend towards spontaneous fission. Indeed, as Seaborg and his colleagues strove to make heavier and heavier elements (they created nine new ones over the next twenty years, and element 106 is now named seaborgium in his honor), they found that these were increasingly unstable, some of them breaking up within microseconds of being made. There seemed good grounds for supposing that one might never get beyond element 108—that this would be the absolute "ultimium."

. . .

THEN, IN THE LATE 1960S, a radical new concept of the nucleus emerged—the notion that its protons and neutrons were arranged in "shells" (like the "shells" of electrons that whirled around the nucleus). The stability of the nucleus of an atom, it was theorized, depended on whether these nuclear shells were filled, just as the chemical stability of atoms depended on the filling of their electron shells. It was calculated that the ideal (or "magic") number of protons required to fill such a nuclear shell would be 114, and the ideal number of neutrons would be 184. A nucleus with both these numbers, a "doubly magic" nucleus, might be, despite its enormous size, remarkably stable.

This idea was startling, paradoxical—as strange and exciting as that of black holes or dark energy. It moved even sober scientists like Seaborg to allegorical language. He thus spoke of a sea of instability—the increasingly and sometimes fantastically unstable elements from 101 to 111—that one would somehow have to leap over if one was ever to reach what he called the island of stability (an elongated island stretching from elements 112 to 118 but having in its center the "doubly magic" isotope of 114). The term "magic" was continually used—Seaborg and others spoke of a magic ridge, a magic mountain, a magic island of elements.

This vision came to haunt the imagination of physicists the world over. Whether or not it was scientifically important, it became psychologically imperative to reach, or at least to sight, this magic territory. There were undertones of other allegories as well—the island of stability could be seen as a topsy-turvy, Alice-in-Wonderland realm where bizarre and gigantic atoms lived their strange lives. Or, more wistfully, the island of stability could be imagined as a sort of Ithaca, where the atomic wanderer, after decades of struggle in the sea of instability, might reach a final haven.

NO EFFORT OR EXPENSE was spared in this enterprise. The vast atom smashers, the particle colliders of Berkeley, Dubna, and Darmstadt, were all enlisted in the quest, and scores of brilliant workers devoted their lives to it. Finally, in 1998, after more than thirty years, the work paid off. Scientists reached the outlying shores of the magic island: they were able to create an isotope of 114, albeit nine neutrons short of the magic number. (When I met Glenn Seaborg in December 1997, he said that one of his longest-lasting and most cherished dreams was to see one of these magic elements—but, sadly, when the creation of 114 was announced in 1999, Seaborg had been disabled by a stroke, and may never have known that his dream had been realized.)

Since elements in the vertical groups of the periodic table are analogous to one another, one can say with confidence that one of the new elements, 113, is a heavier analogue of element 81, thallium. Thallium, a heavy, soft, leadlike metal, is one of the most peculiar of elements, with chemical properties so wild and contradictory that early chemists did not know where to place it in the periodic table. It was sometimes called the platypus of elements. Is thallium's new, heavier analogue, "super-thallium," as strange?

SIMILARLY, the other new element, 115, is certain to be a heavier analogue of element 83, bismuth. As I write, I have a lump of bismuth in front of me, prismatic and terraced like a miniature

Hopi village, glittering with iridescent oxidation colors, and I cannot help wondering whether "super-bismuth," if it could be obtained in massive form, would be as beautiful—or perhaps more so.

And it could be possible to obtain more than a few atoms of these superheavy elements, for they may have half-lives of many years, unlike the elements preceding them, which vanish in split seconds. Atoms of element 111, the heavier analogue of gold, break down in less than a millisecond, and it is difficult to have more than an atom or two at a time, so we may never hope to see what "super-gold" looks like. But if we can make isotopes of 113, 114 (super-lead), and 115, which may have half-lives of years or centuries, we will have three enormously dense and strange new metals.

Of course, we can only guess at what properties 113 and 115 will possess. One can never tell in advance what the practical use or scientific implications of anything new might be. Who would have thought that germanium—an obscure "semimetal" discovered in the 1880s—would turn out to be crucial to the development of transistors? Or that elements like neodymium and samarium, regarded for a century as mere curiosities, would be essential to the making of unprecedentedly powerful permanent magnets?

Such questions are, in a sense, beside the point. We search for the island of stability because, like Mount Everest, it is there. But, as with Everest, there is profound emotion, too, infusing the scientific search to test a hypothesis. The quest for the magic island shows us that science is far from being coldness and calculation, as many people imagine, but is shot through with passion, longing, and romance.

# Reading the Fine Print

I have just had a new book published, but I am unable to read it because, like millions of others, I have impaired vision. I need to use a magnifying glass, and this is cumbersome and slow because the field is restricted and one cannot take in a whole line, let alone a paragraph, at a glance. What I really need is a large-print edition, one that I can read (in bed or in the bath, where I do most of my reading) like any other book. Some of my earlier books existed in large-print editions, invaluable when I was asked to give a public reading. Now I am told a printed version is not "necessary"; instead, we have e-books, which allow us to blow up the size of the type as much as we want.

But I do not want a Kindle or a Nook or an iPad, any one of which could be dropped in the bath or broken, and which has controls I would need a magnifying glass to see. I want a real book made of paper with print—a book with heft, with a bookish smell, as books have had for the last 550 years, a book that I can slip into my pocket or keep with its fellows on my bookshelves, where my eye might alight on it at unexpected times.

When I was a boy, some of my elderly relatives, as well as a young cousin with poor eyesight, used magnifying glasses for reading. The introduction of large-print books in the 1960s was a great boon for them, as it was for all partly sighted readers. Publishing firms that specialized in large-print editions for libraries, schools, and individual readers sprang up, and one could always find these in bookshops or libraries.

In January of 2006, when my vision began to decline, I wondered what I would do. There were audiobooks—I had recorded some of them myself—but I was quintessentially a reader, not a listener. I have been an inveterate reader as far back as I can remember—I often hold page numbers or the look of paragraphs and pages in my mind almost automatically, and I can instantly find my way to a particular passage in most of my books. I want books that belong to me, books whose intimate pagination will become dear and familiar. My brain is geared towards reading— and the answer, for me, clearly, is large-print books.

But one is hard put now to find any large-print books of quality in a bookshop. This I discovered when I went recently to the Strand, a bookshop famous for its miles of shelves, to which I have been going for fifty years. They did have a (small) large-print section, but it consisted mostly of how-to books and trashy novels. There were no collections of poetry, no plays, no biographies, no science. No Dickens, no Jane Austen, none of the classics—no Bellow, no Roth, no Sontag. I came out frustrated, and furious: did publishers think the visually impaired were intellectually impaired, too?

Reading is a hugely complex task, one that calls upon many parts of the brain, but it is not a skill humans have acquired through evolution (unlike speech, which is largely hardwired). Reading is a relatively recent development, arising perhaps five thousand years ago, and it depends on a tiny area of the brain's visual cortex. What we now call the visual word form area is part of a cortical region near the back of the left side of the brain that evolved to recognize basic shapes in nature but can be redeployed for the recognition of letters or words. This elementary shape or letter recognition is only the first step.

From this visual word form area, two-way connections must be made to many other parts of the brain, including those responsible for grammar, memories, association, and feelings, so that letters and words acquire their particular meanings for us. We each form unique neural pathways associated with reading, and we each bring to the act of reading a unique combination not only of memory and experience, but of sensory modalities, too. Some people may "hear" the sounds of the words as they read (I do, but only if I am reading for pleasure, not when I am reading for information); others may visualize them, consciously or not. Some may be acutely aware of the acoustic rhythms or emphases of a sentence; others are more aware of its look or its shape.

In my book *The Mind's Eye,* I describe two patients, both gifted writers, who each lost the ability to read as a result of brain damage to the visual word form area (patients with this sort of alexia can write but cannot read what they have written). One of them, Charles Scribner, Jr., though a publisher himself and a lover of print, turned at once to audiobooks for "reading," and he began dictating, rather than writing, his own books. He found the transition easy—indeed, it seemed to occur by itself. The other man, a crime novelist, Howard Engel, was too deeply rooted in reading and writing to give them up. He continued to write (rather than dictate) his subsequent books, and to find, or devise, an extraordinary new way of "reading"—his tongue started to copy the words in front of him, tracing them on the back of his teeth—he was reading, in effect, by writing with his tongue, employing the motor and tactile areas of his cortex. This, too, seemed to occur by itself. Each man's brain, using its unique strengths and experiences, found the right solution, the right adaptation to the loss.

For someone who is born blind, with no visual imagery at all, reading may be essentially a tactile experience, through the raised print of braille. Braille books, like large-print books, are less and less available now, as people turn to the cheaper and more readily available audiobooks or computer voice programs. But there is a fundamental difference between reading and being read to. When one reads actively, whether using the eyes or a finger, one is free to skip ahead or back, to reread, to ponder or daydream in the middle of a sentence—one reads in one's own time. Being read to, listening to an audiobook, is a more passive experience, subject to the vagaries of another's voice and largely unfolding in the narrator's own time.

If we are forced, later in life, to learn new ways of reading—to accommodate a loss of vision, for instance—we must each adapt in our own way. Some of us may turn from reading to listening; others will continue reading as long as possible. Some may enlarge print on their e-book readers, others on their computers. I have never adopted either of these technologies; for now, at least, I am sticking to the old-fashioned magnifying glass (I have a dozen, in different shapes and strengths).

Writing should be accessible in as many formats as possible— George Bernard Shaw called books the memory of the race. No one sort of book should be allowed to disappear, for we are all individuals, with highly individualized needs and preferences—preferences embedded in our brains at every level, our individual neural patterns and networks creating a deeply personal engagement between author and reader.

# The Elephant's Gait

A recent issue of the science journal *Nature* had a fascinating article, by John Hutchinson and others, entitled
"Are Fast-Moving Elephants Really Running?" The elephants
tested—there were forty-two of them—were marked with paint
dots over their shoulder, hip, and limb joints and videotaped as
they moved along a thirty-meter course (they had ten meters
at each end to accelerate and slow down). It was clear that at
high speed there was an abrupt change of gait, but it was not
so easy to interpret this. Should their rapid shuffle be regarded
as "running"?

Seeing a photo of a marked-up elephant made me think of
how Étienne-Jules Marey, a hundred and fifteen years earlier,
had made a pioneer investigation of elephant gaits, using not
video analysis, of course, but still photography, and marking up
his elephants in much the same fashion. I had just read a book
on Marey, as it happened—a marvelous book by Marta Braun,
entitled *Picturing Time*—along with Rebecca Solnit's acclaimed
biography of Eadweard Muybridge, *River of Shadows*.

Marey and Muybridge were exact contemporaries—they
were both born and died within a few weeks of each other. They

also shared the same initials, EJM, but were otherwise about as different as could be. Muybridge was impulsive, flamboyant, a brilliant peripatetic artist and photographer, drawn in many different creative directions, while Marey was quiet, modest, concentrated, and systematic, spending his entire creative lifetime in his physiology laboratory. And yet, for a brief and crucial time, their lives came together, their ideas interacted, and with this a revolution occurred that not only paved the way for the development of cinematography but created a new tool for science, for the study of time, and for the representation of time and motion in art.

The name of Muybridge is widely known—he is almost an American icon—but Marey has been all but forgotten, though he was famous in his lifetime. Marey's legacy is in many ways richer than Muybridge's, but it was essentially the conjunction of the two men that brought about the great change. Neither alone could have achieved it.

Marey's lifelong fascination with movement started with the internal movements and processes of the body. He had been a pioneer here, inventing pulse meters, blood-pressure graphings, and heart tracings—ingenious precursors of the mechanical instruments we still use in medicine today. Then, in 1867, he moved to the analysis of animal and human locomotion. He used pressure gauges, rubber tubes, and graphic recordings to measure the movements and positions of the limbs, as well as the forces they exerted, when a horse galloped or trotted. From these recordings he made drawings, and these he rotated in a zoetrope, reconstructing in slow motion the movements of the horse.

It had never occurred to him, apparently, to use photography—this, it must have seemed to him and all his contemporaries, was technically impossible. Cameras at this time had no shutters; one still had to remove the lens cap and replace it by hand, so exposures of much less than a second were impossible. Photographic emulsions were not too sensitive, so that an exposure of much less than a second, even if mechanically possible, might fail to admit enough light to create an image on the desperately slow wet plates then in use. And even if one were, somehow, to obtain a single "instantaneous" photograph, how could one obtain ten or twenty in a single second, when each photographic plate took several minutes to develop?

Muybridge, on the other hand, a gifted photographer, had had no special interest in animal movement prior to the 1870s, though he was, as Solnit brings out, always haunted by a sense of the ephemeral, a need to "fix," photographically, the fugitive and the transient (this had earlier led him to make studies of the incessantly changing patterns of clouds). It was only when he met the immensely wealthy railroad baron Leland Stanford, who owned a large racing stable, that Muybridge's future career was determined.

Racing men often debated among themselves as to whether, in galloping, a horse ever had all four hooves off the ground at the same moment—Stanford himself had taken a large bet on this, and he commissioned Muybridge to secure a photo of a horse in mid-gallop if he could. To do this, Muybridge had to make great technical advances, developing faster emulsions and designing shutters that could give an exposure of as little as $\frac{1}{200}$th of a second. Having done so, he produced, in 1873, a single instantaneous photograph of a horse that showed (though not quite as convincingly as Stanford would have liked, for it was not much more than a blurred silhouette) all four hooves indeed suspended in midair.

This might have been the end of the matter, had Stanford not received and excitedly read, at this juncture, Marey's just-published *Animal Mechanism: A Treatise on Terrestrial and Aerial Locomotion*. Here Marey described in exhaustive detail his mechanical and pneumatic means of recording animal motion, showing the sequence of drawings he had constructed from his measurements and how he could bring these to life with the use of a zoetrope. (One of his drawings showed a galloping horse in midair, all of its hooves apparently off the ground.) Stanford saw in a flash that all the postures and movements of the horse as it galloped and trotted could, in principle, be captured photographically in this way, and that one might achieve the miracle of picturing motion—and this, he told Muybridge, was what he must do.

Muybridge, a superb and inventive photographer (his extraordinary pictures of Yosemite, taken with a vast wet-plate camera from the most unexpected angles and viewpoints, are still unmatched today), saw at once that the challenge was to get the horse to take its own photographs. The brilliant notion he conceived and finally brought to perfection was to set up a series of twelve (and later twenty-four) cameras along a measured track, where their shutters would be tripped in quick-fire sequence by the horse as it galloped past. Finally, in 1878, after four years of experiment, he was able to publish his legendary serial photographs. Nothing like these had ever been seen before. Artists had tried to represent the postures of galloping horses for hundreds of years, but with indifferent success and little agreement, for the movements of a galloping horse were too fast for the eye to take in.

Marey, still locked into his own laborious methods, after eleven years of experiments, was stunned when he saw a reproduction of Muybridge's photographs in a magazine, and wrote an urgent and admiring letter to the editor of the magazine: "I am filled with admiration for Mr. Muybridge's instantaneous photographs," he wrote. "Could you put me in touch with [him]?" He imagined a collaboration with Muybridge that would eventuate in seeing "all imaginable animals in their true paces . . . animated zoology." And he foresaw, as Muybridge did, that such photographs could be "for artists . . . a revolution, since they will be provided with the true attitudes of movement, those positions of the body in unstable balance for which no model can pose. You see," he concluded, "my enthusiasm is boundless."

Muybridge responded with equal generosity and grace, telling Marey that his "celebrated work on animal movement first inspired . . . the idea . . . of resolving the problem of locomotion with the help of photography." The two of them subsequently met, cordially, in Paris.

Marey, guided by his previous method of graphic representation—"kymograms," which overlaid, in diagrammatic form, the successive positions of joints and limbs in motion—now devised a photographic parallel to it. Using a single camera with its lens open, he positioned behind the lens a slotted metal disk, which would rotate and serve as a shutter, allowing him to get a dozen or more exposures superimposed on a single plate. These composites, which compressed time into a single frame, Marey called "chronophotographs," and they not only were visually striking (a famous early example was a plate showing the successive positions of a cat as it rotated and righted itself while falling to the ground), but also permitted, as Muybridge's separate frames did not, accurate visualization and analysis of the biomechanics involved.

Towards the end of the 1880s, with the development of flexible celluloid film, both Muybridge and Marey went on to develop cine-cameras, though neither was interested in "cinema" as such, but rather, as Braun puts it, in "capturing the invisible rather than reconstituting the visible."

Marey, with his chronophotographs, went on to study gymnasts and other athletes, workers on assembly lines, and the movement and forces of air and water (he was the first to make wind tunnels) and to pioneer time-lapse underwater photography, which could make the almost invisibly slow movements of sea urchins visible and intelligible. Muybridge focused more on the representation of social interaction and gesture. Both, however, retained their love for "animated zoology," and both, in the mid-1880s, photographed elephants in motion.

Muybridge went back to the technique he had developed on Stanford's farm, using a battery of twenty-four cameras. Marey, however, using his slot-shuttered "photographic gun," and marking his elephants with pieces of paper on their joints, was able to get all the stages of the elephant's movement on a single plate, in a series of overlapping, ghostly images, which showed the vertical movement of the shoulder and hip joints. Such composites give one an extraordinary sense of movement, of an elephant's actual motion and the intricate mechanics involved, which Muybridge's rather static images do not begin to convey. It was Marey's 1887 chronophotograph that rushed into my mind when I read the 2003 *Nature* article on whether elephants run.

The 2003 analysis makes use of sophisticated timers, digitization, and computer analysis—refinements not available in 1887—and with these, it can be shown that elephants in a hurry do, in fact, run and walk simultaneously. That is, the vertical movement of the shoulders indicates a walking motion, while the vertical movement of the hips indicates a running one. It has to be, one guesses, relatively fast walking and relatively slow running; otherwise the hindquarters would collide with the front end. Marey and Muybridge, one feels, would be pleased by this.

# Orangutan

Some years ago while visiting the Toronto Zoo, I visited an orangutan. She was nursing a baby—but when I pressed my bearded face against the window of her large, grassy enclosure, she put her infant down gently, came over to the window, and pressed her face, her nose, opposite mine, on the other side of the glass. I suspect my eyes were darting about as I gazed at her face, but I was much more conscious of *her* eyes. Her bright little eyes—were they orange too?—flicked about, observing my nose, my chin, all the human but also apish features of my face, identifying me (I could not help feeling) as one of her own kind, or at least closely akin. Then she stared into my eyes, and I into hers, like lovers gazing into each other's eyes, with just the pane of glass between us.

I put my left hand against the window, and she immediately put her right hand over mine. Their affinity was obvious—we could both see how similar they were. I found this astounding, wonderful; it gave me an intense feeling of kinship and closeness as I had never had before with any animal. "See," her action said, "my hand, too, is just like yours." But it was also a greeting, like shaking hands or matching palms in a high five.

Then we pulled our faces away from the glass, and she went back to her baby.

I have had and loved dogs and other animals, but I have never known such an instant, mutual recognition and sense of kinship as I had with this fellow primate.

# Why We Need Gardens

As a writer, I find gardens essential to the creative process; as a physician, I take my patients to gardens whenever possible. All of us have had the experience of wandering through a lush garden or a timeless desert, walking by a river or an ocean, or climbing a mountain and finding ourselves simultaneously calmed and reinvigorated, engaged in mind, refreshed in body and spirit. The importance of these physiological states on individual and community health is fundamental and wide-ranging. In forty years of medical practice, I have found only two types of non-pharmaceutical "therapy" to be vitally important for patients with chronic neurological diseases: music and gardens.

The wonder of gardens was introduced to me very early, before the war, when my mother or Auntie Len would take me to the great botanical garden at Kew. We had common ferns in our garden, but not the gold and silver ferns, the water ferns, the filmy ferns, the tree ferns I first saw at Kew. It was at Kew that I saw the gigantic leaf of the great Amazon water lily, *Victoria regia,* and like many children of my era, I was sat upon one of these giant lily pads as a baby.

As a student at Oxford, I discovered with delight a very

different garden—the Oxford Botanic Garden, one of the first walled gardens established in Europe. It pleased me to think that Boyle, Hooke, Willis, and other Oxford figures might have walked and meditated there in the seventeenth century.

I try to visit botanical gardens wherever I travel, seeing them as reflections of their times and cultures, no less than living museums or libraries of plants. I felt this strongly in the beautiful seventeenth-century Hortus Botanicus in Amsterdam, coeval with its neighbor, the great Portuguese Synagogue, and liked to imagine how Spinoza might have enjoyed the former after he had been excommunicated by the latter—was his vision of *"Deus sive Natura"* in part inspired by the Hortus?

The botanical garden in Padua is even older, going right back to the 1540s, and medieval in its design. Here Europeans got their first look at plants from the Americas and the Orient, plant forms stranger than anything they had ever seen or dreamed of. It was here, too, that Goethe, looking at a palm, conceived his theory of the metamorphoses of plants.

When I travel with fellow swimmers and divers to the Cayman Islands, to Curaçao, to Cuba, wherever—I seek out botanical gardens, counterpoints to the exquisite underwater gardens I see when I snorkel or scuba above them.

I HAVE LIVED in New York City for fifty years, and living here is sometimes made bearable for me only by its gardens. This has been true for my patients, too. When I worked at Beth Abraham, a hospital just across the road from the New York Botanical Garden, I found that there was nothing long-shut-in patients loved more than a visit to the garden—they spoke of the hospital and the garden as two different worlds.

I cannot say exactly how nature exerts its calming and organizing effects on our brains, but I have seen in my patients the restorative and healing powers of nature and gardens, even for those who are deeply disabled neurologically. In many cases, gardens and nature are more powerful than any medication.

My friend Lowell has moderately severe Tourette's syndrome: in his usual busy, city environment, he has hundreds of tics and verbal ejaculations each day—grunting, jumping, touching things compulsively. I was therefore amazed one day when we were hiking in a desert to realize that his tics had completely disappeared. The remoteness and uncrowdedness of the scene, combined with some ineffable calming effect of nature, served to defuse his ticcing, to "normalize" his neurological state, at least for a time.

An elderly lady with Parkinson's disease, whom I met in Guam, often found herself frozen, unable to initiate movement—a common problem for those with parkinsonism. But once we led her out into the garden, where plants and a rock garden provided a varied landscape, she was galvanized by this, and could rapidly, unaided, climb up the rocks and down again.

I have a number of patients with very advanced dementia or Alzheimer's disease, who may have very little sense of orientation to their surroundings. They have forgotten, or cannot access, how to tie their shoes or handle cooking implements. But put them in front of a flower bed with some seedlings, and they will know exactly what to do—I have never seen such a patient plant something upside down.

My patients often live in nursing homes or chronic-care institutions, so the physical environment of these settings is crucial in promoting their well-being. Some of these institutions have actively used the design and management of their open spaces to promote better health for their patients. For example, Beth Abraham hospital, in the Bronx, is where I saw the severely parkinsonian postencephalitic patients I wrote about in *Awakenings*. In the 1960s, it was a pavilion surrounded by large gardens. As it expanded to a five-hundred-bed institution, it swallowed most of the gardens, but it did retain a central patio full of potted plants that remains very crucial for the patients. There are also raised beds so that blind patients can touch and smell and wheelchair patients can have direct contact with the plants.

I work also with the Little Sisters of the Poor, who have nursing homes around the world. They are a Catholic order, originally founded in Brittany in the late 1830s, that spread to America in the 1860s. At that time, it was common for an institution such as a nursing home or a state hospital to have a large farm garden and often a dairy as well. Alas, this is a tradition that has mostly vanished, but the Little Sisters are trying to reintroduce it today. One of their New York City residences is situated in a leafy suburb in Queens, with plenty of walkways and benches. Some of the residents can walk by themselves, some need a stick, some need a walker, and some have to be wheeled—but nearly all of them, when it becomes warm enough, want to be outside in the garden and the fresh air.

Clearly, nature calls to something very deep in us. Biophilia, the love of nature and living things, is an essential part of the human condition. Hortophilia, the desire to interact with, manage, and tend nature, is also deeply instilled in us. The role that nature plays in health and healing becomes even more critical for people working long days in windowless offices, for those living in city neighborhoods without access to green spaces, for children in city schools, or for those in institutional settings such as nursing homes. The effects of nature's qualities on health are not only spiritual and emotional but physical and neurological. I have no doubt that they reflect deep changes in the brain's physiology, and perhaps even its structure.

# Night of the Ginkgo

Today in New York—November 13th—leaves are falling, drifting, skittering everywhere. But there is one striking exception: the fan-shaped leaves of the ginkgo are still firmly attached to their branches, even though many of them have turned a luminous gold. One sees why this beautiful tree has been revered since ancient times. Carefully preserved for millennia in the temple gardens of China, ginkgoes are almost extinct in the wild, but they have an extraordinary ability to survive the heat, the snows, the hurricanes, the diesel fumes, and the other charms of New York City, and there are thousands of them here, mature ones bearing a hundred thousand leaves or more—tough, heavy Mesozoic leaves such as the dinosaurs ate. The ginkgo family has been around since before the dinosaurs, and its only remaining member, *Ginkgo biloba,* is a living fossil, basically unchanged in two hundred million years.

While the leaves of the more modern angiosperms—maples, oaks, beeches, what have you—are shed over a period of weeks after turning dry and brown, the ginkgo, a gymnosperm, drops its leaves all at once. The botanist Peter Crane, in his book *Ginkgo,* writes about a very large ginkgo in Michigan, "for many years there was a competition to guess the date on which the leaves would fall." In general, Crane says, it happens with "eerie synchronicity," and he quotes the poet Howard Nemerov:

> *Late in November, on a single night*
> *Not even near to freezing, the ginkgo trees*
> *That stand along the walk drop all their leaves*
> *In one consent, and neither to rain nor to wind*
> *But as though to time alone: the golden and green*
> *Leaves litter the lawn today, that yesterday*
> *Had spread aloft their fluttering fans of light.*

Are the ginkgoes responding to some external signal, such as the change of temperature or light? Or to some internal, genetically programmed signal? No one knows what lies behind this synchronicity, but it is surely related to the antiquity of the ginkgo, which has evolved along a very different path from that of more modern trees.

Will it be November 20th, 25th, 30th? Whenever it is, each tree will have its own Night of the Ginkgo. Few people will see this—most of us will be asleep—but in the morning the ground beneath the ginkgo will be carpeted with thousands of heavy, golden, fan-shaped leaves.

# Filter Fish

Gefilte fish is not an everyday dish; it is to be eaten mainly on the Jewish Sabbath in Orthodox households, when cooking is not allowed. When I was growing up, my mother would take off from her surgical duties early on Friday afternoon and devote her time, before the coming of Shabbat, to preparing gefilte fish and other Sabbath dishes.

Our gefilte fish was basically carp, to which pike, whitefish, and sometimes perch or mullet would be added. (The fishmonger delivered the fish alive, swimming in a pail of water.) The fish had to be skinned, boned, and fed into a grinder; we had a massive metal grinder attached to the kitchen table, and my mother would sometimes let me turn the handle. She would then mix the ground fish with raw eggs, matzo meal, and pepper and sugar. (Litvak gefilte fish, I was told, used more pepper, which is how she made it—my father was a Litvak, born in Lithuania.)

My mother would fashion the mixture into balls about two inches in diameter—two to three pounds of fish would allow a dozen or more substantial fish balls—and then poach these gently with a few slices of carrot. As the gefilte fish cooled, a jelly of an extraordinarily delicate sort coalesced, and, as a child, I had a passion for the fish balls and their rich jelly, along with the obligatory *khreyn* (Yiddish for horseradish).

I thought I would never taste anything like my mother's gefilte fish again, but in my forties I found a housekeeper, Helen Jones, with a veritable genius for cooking. Helen improvised everything, nothing was by the book, and, learning my tastes, she decided to try her hand at gefilte fish.

When she arrived each Thursday morning, we would set out for the Bronx to do some shopping together, our first stop being a fish shop on Lydig Avenue run by two Sicilian brothers who were as like as twins. Though the fishmongers were happy to give us carp, whitefish, and pike, I had no idea how Helen, African-American, a good, churchgoing Christian, would manage with making such a Jewish delicacy. But her powers of improvisation were formidable, and she made magnificent gefilte fish (she called it "filter fish"), which, I had to acknowledge, was as good as my mother's. Helen refined her filter fish each time she made it, and my friends and neighbors got a taste for it, too. So did Helen's church friends; I loved to think of her fellow Baptists gorging on gefilte fish at their church socials.

For my fiftieth birthday, in 1983, she made a gigantic bowl of it—enough for the fifty birthday guests. Among them was Bob Silvers, the editor of *The New York Review of Books*, who was so enamored of Helen's gefilte fish that he wondered if she could make it for his entire staff.

When Helen died, after seventeen years of working for me, I mourned her deeply—and I lost my taste for gefilte fish. Commercially made, bottled gefilte fish, sold in supermarkets, I found detestable compared to Helen's ambrosia.

But now, in what are (barring a miracle) my last weeks

of life—so queasy that I am averse to almost every food and have difficulty swallowing anything except liquids or jellylike solids—I have rediscovered the joys of gefilte fish. I cannot eat more than two or three ounces at a time, but an aliquot of gefilte fish every waking hour nourishes me with much-needed protein. (Gefilte-fish jelly, like calf's-foot jelly, was always valued as an invalid's food.)

Deliveries now arrive daily from one shop or another: Murray's on Broadway, Russ & Daughters, Sable's, Zabar's, Barney Greengrass, the Second Avenue Deli—they all make their own gefilte fish, and I like it all (though none compares to my mother's or Helen's).

While I have conscious memories of gefilte fish from about the age of four, I suspect that I acquired my taste for it even earlier, for, with its abundant, nutritious jelly, it was often given to infants in Orthodox households as they moved from baby foods to solid food. Gefilte fish will usher me out of this life, as it ushered me into it, eighty-two years ago.

# Life Continues

My favorite aunt, Auntie Len, when she was in her eighties, told me that she had not too much difficulty adjusting to all the things that were new in her lifetime—jet planes, space travel, plastics, and so on—but she could not accustom herself to the disappearance of the old. "Where have all the horses gone?" she would sometimes say. Born in 1892, she had grown up in a London full of carriages and horses.

I have similar feelings myself. A few years ago, I was walking with my niece Liz down Mill Lane, a road near the house in London where I grew up. I stopped at a railway bridge where I had loved leaning over the railings as a child. I watched various electric and diesel trains go by, and after a few minutes Liz, getting impatient, asked, "What are you waiting for?" I said I was waiting for a steam train. Liz looked at me as if I were crazy.

"Uncle Oliver," she said. "There haven't been steam trains for more than forty years."

I have not adjusted as well as my aunt to some aspects of the new—perhaps because the rate of social change associated with technological advances has been so rapid and profound. I cannot get used to seeing myriads of people in the street peering into little boxes or holding them in front of their faces, walking blithely in front of moving traffic, totally out of touch with their surroundings. I am most alarmed by such distraction and inattention when I see young parents staring at their cell phones and ignoring their own babies as they walk or wheel them along. Such children, unable to attract their parents' attention, must feel neglected, and they will surely show the effects of this in the years to come.

In his 2007 novel *Exit Ghost,* Philip Roth speaks of how radically changed New York City appears to a reclusive writer who has been away from it for a decade. He is forced to overhear cell-phone conversations all around him, and he wonders, "What had happened in these ten years for there suddenly to be so much to say—so much so pressing that it couldn't wait to be said? . . . I did not see how anyone could believe he was continuing to live a human existence by walking about talking into his phone for half his waking life."

These gadgets, already ominous in 2007, have now immersed us in a virtual reality far denser, more absorbing, and even more dehumanizing.

I am confronted every day by the complete disappearance of the old civilities. Social life, street life, and attention to people and things around one have largely disappeared, at least in big cities, where a majority of the population is now glued almost without pause to their phones or other devices—jabbering, texting, playing computer games, turning more and more to virtual reality of every sort.

Everything is public now, potentially: one's thoughts, one's photos, one's movements, one's purchases. There is no privacy and apparently little desire for it in a world devoted to nonstop use of social media. Every minute, every second, has to be spent with one's device clutched in one's hand. Those trapped in this virtual world are never alone, never able to concentrate and appreciate in their own way, silently. They have given up, to a great extent, the amenities and achievements of civilization: solitude and leisure, the sanction to be oneself, truly absorbed, whether in contemplating a work of art, a scientific theory, a sunset, or the face of one's beloved.

A few years ago, I was invited to join a panel discussion titled "Information and Communication in the Twenty-first Century." One of the panelists, an internet pioneer, said proudly that his young daughter surfed the internet twelve hours a day and had access to a breadth and range of information that no one of a previous generation could have achieved. I asked whether she had read any of Jane Austen's novels, or *any* classic novel, and he said, "No, she doesn't have time for anything like that." I wondered aloud whether she would then have no solid understanding of human nature or society, and suggested that while she might be stocked with wide-ranging information, that was different from knowledge; she would have a mind both shallow and centerless. Half the audience cheered; the other half booed.

Much of this, remarkably, was envisaged by E. M. Forster in his 1909 short story "The Machine Stops," where he imagined a future in which people live underground in isolated cells, never seeing one another and communicating only by audio and visual devices. In this world, original thought and direct observation are discouraged—"Beware of first-hand ideas!" people are told. Humanity has been overtaken by "the Machine," which provides all comforts and meets all needs—except for human contact. One young man, Kuno, pleads with his mother via a Skype-like call, "I want to see you not through the Machine. I want to speak to you not through the wearisome Machine."

He says to his mother, who is absorbed in her hectic, meaningless life, "We have lost the sense of space. . . . We have lost a part of ourselves. . . . Cannot you see . . . that it is we that are dying, and that down here the only thing that really lives is the Machine?"

This is how I feel increasingly often about our bewitched, besotted society, too.

AS ONE'S DEATH GROWS NEAR, one may take comfort in the feeling that life will go on—if not for oneself then for one's children, or for what one has created. Here at least one can invest hope, though there may be no hope for oneself physically and (for those of us who are not believers) no sense of any "spiritual" survival after bodily death.

But it may not be enough to create, to contribute, to have influenced others, if one feels, as I do now, that the very culture in which one was nourished and to which one had given one's best in return is itself threatened.

Though I am supported and stimulated by my friends, by readers around the world, by memories of my life, and by the joy that writing gives me, I have, as many of us must have, deep fears about the well-being and even survival of our world.

Such fears have been expressed at the highest intellectual and moral levels. Martin Rees, Astronomer Royal and former president of the Royal Society, is not a man given to apocalyptic thinking, but in 2003 he published a book called *Our Final*

*Hour,* subtitled *A Scientist's Warning—How Terror, Error, and Environmental Disaster Threaten Humankind's Future in This Century.* More recently, Pope Francis's remarkable encyclical *Laudato Si'* was published, with its deep consideration not only of human-induced climate change and widespread ecological disaster but of the desperate state of the poor and the growing threats of consumerism and misuse of technology. Traditional wars have now been joined by genocide, extremism, and terrorism on a scale never before seen and, in some cases, by the deliberate destruction of our human heritage, of history and culture itself.

These threats of course concern me, but at a distance—I worry more about the subtle, pervasive draining out of meaning, of intimate contact, from our society and culture.

When I was eighteen, I first read Hume and I was horrified by the vision he expressed in his 1738 *Treatise of Human Nature,* in which he wrote that mankind is "nothing but a bundle or collection of different perceptions, which succeed each other with an inconceivable rapidity, and are in a perpetual flux and movement." As a neurologist, I have seen many patients rendered amnesic by destruction of the memory systems in their brains, and I cannot help feeling that these people, having lost any sense of a past or future and caught in a flutter of ephemeral, ever-changing sensations, have in some sense been reduced from human beings to Humean ones.

I have only to venture into the streets of my own neighborhood, the West Village, to see such Humean casualties by the thousand: younger people, for the most part, who have grown up in our social-media era, have no personal memory of how things were before, and no immunity to the seductions of digital life. What we are seeing—and bringing on ourselves—resembles a neurological catastrophe on a gigantic scale.

Nonetheless, I dare to hope that, despite everything, human life and its richness of cultures will survive, even on a ravaged earth. While some see art as a bulwark of our culture, our collective memory, I see science, with its depth of thought, its palpable achievements and potentials, as equally important; and science, good science, is flourishing as never before, though it moves cautiously and slowly, its insights checked by continual self-testing and experiment. Though I revere good writing and art and music, it seems to me that only science, aided by human decency, common sense, farsightedness, and concern for the unfortunate and the poor, offers the world any hope in its present morass. This is explicit in Pope Francis's encyclical and may be practiced not only with gigantic, centralized technologies but by workers, artisans, and farmers in the villages of the world. Between us, we can surely pull the world through its present crises and lead the way to a happier time ahead. As I face my own impending departure from the world, I have to believe in this—that mankind and our planet will survive, that life will continue, and that this will not be our final hour.

# Bibliography

Alexander, Eben. 2012. *Proof of Heaven: A Neurosurgeon's Journey into the Afterlife.* New York: Simon & Schuster.

Braun, Marta. 1992. *Picturing Time: The Work of Etienne-Jules Marey (1830–1904).* Chicago: University of Chicago Press.

Cohen, Donna, and Carl Eisdorfer. 2001. *The Loss of Self: A Family Resource for the Care of Alzheimer's Disease and Related Disorders.* New York: Norton.

Coleridge, Samuel Taylor. *Encyclopaedia Metropolitana* (reprinted in *The Friend* as "Essays as Method").

Crane, Peter. 2013. *Ginkgo: The Tree That Time Forgot.* New Haven: Yale University Press.

Crick, Francis. 1981. *Life Itself: Its Origin and Nature.* New York: Simon & Schuster.

Crick, Francis, and Leslie Orgel. 1973. "Directed Panspermia." *Icarus* 19: 341–46.

Crick, Francis, and Graeme Mitchison. 1983. "The Function of Dream Sleep." *Nature* 304 (5922): 111–14.

Custance, John. 1952. *Wisdom, Madness and Folly: The Philosophy of a Lunatic.* New York: Pellegrini.

Davy, Humphry. 1813. *Elements of Agricultural Chemistry in a Course of Lectures.* London: Longman.

———. 1817. "Some Researches on Flame." *Philosophical Transactions of the Royal Society of London* 107: 145–76.

———. 1828. *Salmonia; or Days of Fly Fishing.* London: John Murray.

Dawkins, Richard. 1996. *Climbing Mount Improbable.* New York: Norton.

DeBaggio, Thomas. 2002. *Losing My Mind: An Intimate Look at Life with Alzheimer's.* New York: Free Press.

———. 2003. *When It Gets Dark: An Enlightened Reflection on Life with Alzheimer's*. New York: Free Press.

de Duve, Christian. 1995. *Vital Dust: Life as a Cosmic Imperative*. New York: Basic Books.

Dewhurst, Kenneth, and A. W. Beard. 1970. "Sudden Religious Conversions in Temporal Lobe Epilepsy." *British Journal of Psychiatry* 117: 497–507.

Dyson, Freeman J. 1999. *Origins of Life*. Second edition. Cambridge: Cambridge University Press.

Edelman, Gerald M. 1987. *Neural Darwinism: The Theory of Neuronal Group Selection*. New York: Basic Books.

Ehrsson, H. Henrik, Charles Spence, and Richard E. Passingham. 2004. "That's My Hand! Activity in the Premotor Cortex Reflects Feeling of Ownership of a Limb." *Science* 305 (5685): 875–77.

Ehrsson, H. Henrik, Nicholas P. Holmes, and Richard E. Passingham. 2005. "Touching a Rubber Hand: Feeling of Body Ownership is Associated with Activity in Multisensory Brain Areas." *Journal of Neuroscience* 25 (45): 10564–73.

Ehrsson, H. Henrik. 2007. "The Experimental Induction of Out-of-Body Experiences." *Science* 317 (5841): 1048.

Erikson, Erik, Joan Erikson, and Helen Kivnick. 1987. *Vital Involvement in Old Age*. New York: Norton.

Forster, E. M. 1909/1928. "The Machine Stops." In *The Eternal Moment*. London: Sidgwick and Jackson.

Freud, Sigmund. 1900. *Interpretation of Dreams*. Standard edition, 5.

Gajdusek, Carleton. 1989. "Fantasy of a 'Virus' from the Inorganic World." *Haematology and Blood Transfusion* 32 (February): 481–99.

Goffman, Erving. 1961. *Asylums: Essays on the Social Situation of Mental Patients and Other Inmates*. New York: Anchor.

Goldstein, Kurt. 1934/2000. *The Organism*. With a foreword by Oliver Sacks. New York: Zone Books.

Gould, Stephen Jay. 1985. *The Flamingo's Smile: Reflections in Natural History*. New York: Norton.

Gray, Spalding. 2012. *The Journals of Spalding Gray*. Edited by Nell Casey. New York: Vintage.

Greenberg, Michael. 2008. *Hurry Down Sunshine*. New York: Other Press.

Groopman, Jerome. 2007. *How Doctors Think*. New York: Houghton Mifflin.

Hobbes, Thomas. 1651/1904. *Leviathan*. Cambridge: Cambridge University Press.

Holmes, Richard. 1989. *Coleridge: Early Visions, 1772–1804*. New York: Pantheon.

Hoyle, Fred, and Chandra Wickramasinghe. 1982. *Evolution from Space: A Theory of Cosmic Creationism*. New York: Simon & Schuster.

Humboldt, Alexander von. 1845/1997. *Cosmos*. Baltimore, Md.: Johns Hopkins University Press.

Hume, David. 1738/1874. *Treatise of Human Nature*. London: Longmans, Green.

Hutchinson, John, Dan Famini, Richard Lair, and Rodger Kram. 2003. "Biomechanics: Are Fast-Moving Elephants Really Running?" *Nature* 422: 493–94.

Ibsen, Henrik. 1888/2001. *The Lady from the Sea*. In *Four Major Plays*, vol. 2. Translated and with a foreword by Rolf Fjelde. New York: Signet Classics.

Jackson, J. Hughlings. 1894/2001. "The Factors of Insanities." Classic Text No. 47. *History of Psychiatry* 12 (47): 353–73.

Jamison, Kay Redfield. 1993. *Touched with Fire: Manic-Depressive Illness and the Artistic Temperament*. New York: Free Press.

———. 1995. *An Unquiet Mind: A Memoir of Moods and Madness*. New York: Knopf.

Jelliffe, Smith Ely. 1927. *Post-Encephalitic Respiratory Disorders*. Washington, DC: Nervous and Mental Disease Publishing Co.

Joyce, James. 1922. *Finnegans Wake*. London: Faber and Faber.

Karinthy, Frigyes. 1939/2008. *A Journey Round My Skull*. With an introduction by Oliver Sacks. New York: New York Review Books.

King, Lucy. 2002. *From Under the Cloud at Seven Steeples, 1878–1885: The Peculiarly Saddened Life of Anna Agnew at the Indiana Hospital for the Insane*. Zionsville: Guild Press of Indiana.

Knight, David. 1992. *Humphry Davy: Science and Power*. Cambridge: Cambridge University Press.

Kurlan, R., J. Behr, L. Medved, I. Shoulson, D. Pauls, J. Kidd, K. K. Kidd. 1986. "Familial Tourette Syndrome: Report of a Large Pedigree and Potential for Linkage Analysis." *Neurology* 36: 772–76.

Liveing, Edward. 1873. *On Megrim, Sick-Headache, and Some Allied Disorders: A Contribution to the Pathology of Nerve-Storms*. London: Churchill.

Lowell, Robert. 1959. Draft manuscript for *Life Studies*. Houghton Library, Harvard College Library.

Luhrmann, T. M. 2012. *When God Talks Back: Understanding the American Evangelical Relationship with God*. New York: Knopf.

Marey, E. J. 1879. *Animal Mechanism: A Treatise on Terrestrial and Aerial Locomotion.* New York: Appleton.

Margulis, Lynn, and Dorion Sagan. 1986. *Microcosmos: Four Billion Years of Microbial Evolution.* New York: Summit Books.

Mayr, Ernst. 1997. *This Is Biology: The Science of the Living World.* Cambridge, Mass.: Belknap Press of Harvard University Press.

Merzenich, Michael. 1998. "Long-term Change of Mind." *Science* 282 (5391): 1062–63.

Monod, Jacques. 1971. *Chance and Necessity: An Essay on the Natural Philosophy of Modern Biology.* New York: Knopf.

Nelson, Kevin. 2011. *The Spiritual Doorway in the Brain: A Neurologist's Search for the God Experience.* New York: Dutton.

Neugeboren, Jay. 1997. *Imagining Robert: My Brother, Madness, and Survival.* New York: Morrow.

———. 2008. "Infiltrating the Enemy of the Mind." Review of *The Center Cannot Hold,* by Elyn Saks. *New York Review of Books,* April 17.

Parks, Tim. 2000. "In the Locked Ward." Review of *Imagining Robert,* by Jay Neugeboren. *New York Review of Books,* February 24.

Payne, Christopher. 2009. *Asylum: Inside the Closed World of State Mental Hospitals.* With a foreword by Oliver Sacks. Cambridge, Mass.: MIT Press.

Penney, Darby, and Peter Stastny. 2008. *The Lives They Left Behind: Suitcases from a State Hospital Attic.* New York: Bellevue Literary Press.

Podvoll, Edward M. 1990. *The Seduction of Madness: Revolutionary Insights into the World of Psychosis and a Compassionate Approach to Recovery at Home.* New York: HarperCollins.

Provine, Robert. 2012. *Curious Behavior: Yawning, Laughing, Hiccupping, and Beyond.* Cambridge, Mass.: Belknap Press of Harvard University Press.

Rees, Martin. 2003. *Our Final Hour: A Scientist's Warning—How Terror, Error, and Environmental Disaster Threaten Humankind's Future in This Century.* New York: Basic Books.

Rhodes, Richard. 1997. *Deadly Feasts: Tracking the Secrets of a Terrifying New Plague.* New York: Simon & Schuster.

Roosens, Eugeen. 1979. *Mental Patients in Town Life: Geel—Europe's First Therapeutic Community.* Beverly Hills: Sage Publications.

Roosens, Eugeen, and Lieve Van de Walle. 2007. *Geel Revisited: After Centuries of Mental Rehabilitation.* Antwerp: Garant.

Roth, Philip. 2007. *Exit Ghost.* New York: Houghton Mifflin Harcourt.

Sacks, Oliver. 1973. *Awakenings.* New York: Doubleday.

———. 1984. *A Leg to Stand On.* New York: Summit.

———. 1985. *The Man Who Mistook His Wife for a Hat.* New York: Summit.

———. 1992. *Migraine.* Rev. ed. New York: Vintage.

———. 1995. *An Anthropologist on Mars.* New York: Knopf.

———. 2001. *Uncle Tungsten.* New York: Knopf.

———. 2007. *Musicophilia: Tales of Music and the Brain.* New York: Knopf.

———. 2010. *The Mind's Eye.* New York: Knopf.

———. 2012. *Hallucinations.* New York: Knopf.

———. 2015. *On the Move.* New York: Knopf.

Saks, Elyn. 2007. *The Center Cannot Hold: My Journey Through Madness.* New York: Hyperion.

Sebald, W. G. 1998. *The Rings of Saturn.* New York: New Directions.

Sheehan, Susan. 1982. *Is There No Place on Earth for Me?* New York: Houghton Mifflin Harcourt.

Shelley, Mary. 1818. *Frankenstein; or, The Modern Prometheus.* London: Lackington, Hughes, Harding, Mavor & Jones.

Shengold, Leonard. 1993. *The Boy Will Come to Nothing! Freud's Ego Ideal and Freud as Ego Ideal.* New Haven: Yale University Press.

Shubin, Neil. 2008. *Your Inner Fish: A Journey into the 3.5-Billion-Year History of the Human Body.* New York: Pantheon.

Smylie, Mike. 2004. Herring: A History of the Silver Darlings. Stroud, UK: Tempus.

Solnit, Rebecca. 2003. *River of Shadows: Eadweard Muybridge and the Technological Wild West.* New York: Viking.

Wells, H. G. 1898. *The War of the Worlds.* London: Heinemann.

———. 1901/2003. *The First Men in the Moon.* New York: Modern Library.

# Permissions and Acknowledgments

Grateful acknowledgment is made to the following for permission to reprint previously published material:

Alfred A. Knopf, an imprint of the Knopf Doubleday Publishing Group, a division of Penguin Random House LLC: Excerpt of "Into the Sun" from *An Unquiet Mind* by Kay Redfield Jamison, copyright © 1995 by Kay Redfield Jamison. Reprinted by permission of Alfred A. Knopf, an imprint of the Knopf Doubleday Publishing Group, a division of Penguin Random House LLC. All rights reserved.

Farrar, Straus and Giroux: Excerpts from *Wisdom, Madness and Folly: The Philosophy of a Lunatic* by John Custance. Copyright © 1951 by John Custance, copyright renewed 1980 by John Custance. Reprinted by permission of Farrar, Straus and Giroux.

HarperCollins Publishers Ltd. and Other Press, LLC: Excerpt from *Hurry Down Sunshine: A Father's Memoir of Love and Madness* by Michael Greenberg, copyright © 2008 by Michael Greenberg. Reprinted by permission of HarperCollins Publishers Ltd. and Other Press, LLC. Any third party use of this material outside of this publication is prohibited. All rights reserved.

Several works were previously published, some in different form, in the following publications:

### FIRST LOVES

"Water Babies" first appeared in *The New Yorker,* May 26, 1997.

"Remembering South Kensington" first appeared in *Discover,* November 1991.

"First Love" first appeared in *The New York Review of Books,* October 18, 2001, and in *Uncle Tungsten.*

"Humphry Davy: Poet of Chemistry" first appeared in longer form in *The New York Review of Books,* November 4, 1993.

"Libraries" first appeared in *The Threepenny Review,* Fall 2014.

"A Journey Inside the Brain" first appeared in slightly different form in *The New York Review of Books,* March 20, 2008, and as an introduction to Frigyes Karinthy, *A Journey Round My Skull* (New York: New York Review Books, 2008).

## CLINICAL TALES

"Cold Storage" first appeared in slightly different form in *Granta,* Spring 1987.

"Neurological Dreams" first appeared in somewhat different form in *MD* 35, no. 2 (February 1991) and in Deirdre Barrett, ed., *Trauma and Dreams* (Cambridge, Mass.: Harvard University Press, 1996).

"Nothingness" first appeared in slightly different form in Richard L. Gregory, ed., *The Oxford Companion to the Mind* (New York: Oxford University Press, 1987).

"Seeing God in the Third Millennium" first appeared on www.theatlantic .com, December 2012.

"Hiccups and Other Curious Behaviors" is previously unpublished.

"Travels with Lowell" is previously unpublished, and incorporates some of "The Divine Curse," which originally appeared in *Life,* September 1988.

"Urge" first appeared in *The New York Review of Books,* September 24, 2015.

"The Catastrophe" first appeared in *The New Yorker,* April 27, 2015.

"Dangerously Well" is based on an article by Oliver Sacks and Melanie Shulman published in *Neurology* 64 (2005) under the title "Steroid Dementia: An Overlooked Diagnosis?"

"Tea and Toast" is previously unpublished.

"Telling" is previously unpublished.

"The Aging Brain" is based on an article published in the *Archives of Neurology,* October 1997.

## Permissions and Acknowledgments

"Kuru" first appeared in somewhat different form in *The New Yorker,* April 14, 1997, under the title "Eat, Drink, and Be Wary."

"A Summer of Madness" first appeared in *The New York Review of Books,* September 25, 2008.

"The Lost Virtues of the Asylum" first appeared in slightly different form in *The New York Review of Books* (September 24, 2009) and as a foreword in Christopher Payne, *Asylum* (Cambridge, Mass.: MIT Press, 2009).

### LIFE CONTINUES

"Anybody Out There?" first appeared, in slightly different form, in *Natural History,* November 2002, and in *Astrobiology Magazine,* December 2002.

"Clupeophilia" first appeared in *The New Yorker,* July 20, 2009.

"Colorado Springs Revisited" first appeared in *Columbia: A Journal of Literature and Art,* Spring 2010.

"Botanists on Park" first appeared in *The New Yorker,* August 13, 2007.

"Greetings from the Island of Stability" first appeared in *The New York Times,* February 8, 2004.

"Reading the Fine Print" first appeared in *The New York Times Book Review,* December 14, 2012.

"The Elephant's Gait" first appeared in the journal *Omnivore,* Autumn 2003.

"Orangutan" is previously unpublished.

"Why We Need Gardens" is previously unpublished.

"Night of the Ginkgo" first appeared in *The New Yorker,* November 24, 2014.

"Filter Fish" first appeared in *The New Yorker,* September 14, 2015.

"Life Continues" is previously unpublished.

# Index